I0020716

Building Blazor WebAssembly Applications with gRPC

Learn how to implement source generators and gRPC in your Blazor apps for better performance

Václav Pekárek

BIRMINGHAM—MUMBAI

Building Blazor WebAssembly Applications with gRPC

Group Product Manager: Pavan Ramchandani
Publishing Product Manager: Jane D'Souza
Senior Editor: Divya Anne Selvaraj
Technical Editor: Joseph Aloocaran
Copy Editor: Safis Editing
Project Coordinator: Manthan Patel
Proofreader: Safis Editing
Indexer: Rekha Nair
Production Designer: Roshan Kawale
Marketing Coordinator: Anamika Singh

First published: October 2022

Production reference: 1061022

Published by Packt Publishing Ltd.
Livery Place
35 Livery Street
Birmingham
B3 2PB, UK.

ISBN 978-1-80461-055-8

www.packt.com

To my parents, Jaroslava and Antonín, for their sacrifices and for the education they gave me. To my wife, Sylvie, for pushing me further and believing in me.

- Václav Pekárek

Contributors

About the author

Václav Pekárek is an experienced developer with 10 years of web development experience using .NET and C#. He is also a Microsoft Certified Professional.

Václav is the owner of a small IT company in Ostrava, Czech Republic. You can contact him on his LinkedIn page, on Twitter at @vaclavpekarek, or via email at vaclav.pekarek@evolutit.cz.

I want to thank my wife Sylvie for her support during the entire course of writing this book. I also want to thank Divya Anne Selvaraj for her ideas for improvements and for keeping the book on track, and Jane D'Souza for making this happen.

About the reviewer

Eduardo Padilla is a seasoned senior consultant and software architect/developer with more than 20 years of experience using Microsoft technologies. He has worked in different industries (health, finance, travel, legal, logistics, and telecommunications) for Fortune 500 companies.

Eduardo is very enthusiastic about emerging technologies and how to use them to improve existing solutions, which made him look into Blazor and the myriad of possibilities it opened. He became an early adopter, the same way as he did with the .NET implementation of gRPC.

Eduardo provides consulting services through his small company in Miami, FL. Feel free to contact him at `eduardo.padilla@me3technologies.com` or through his LinkedIn page.

Table of Contents

3

Creating a Database Using Entity Framework Core   51

4

Connecting Client and Server with REST API   81

5

Building gRPC Services   103

6

Diving Deep into Source Generators   135

7

Best Practices for C# and gRPC   159

Preface

This book will introduce you to the Blazor WebAssembly technology and help you in deciding to use this technology in your projects. Blazor is an ideal framework for developing the frontend of websites and can compete well with existing JavaScript frameworks.

While there are multiple development options within Blazor, this book is focused on the WebAssembly hosting model, where you can create static websites or dynamic websites with a separate frontend and backend.

The idea of using gRPC instead of REST for the Blazor WebAssembly application proves itself as a great option within shared models, leading to a strongly typed client and server available in both parts of the Blazor WebAssembly application.

In the last part of this book, you will also be introduced to source generators and learn how you can use them to generate repetitive code with zero mistakes.

Who this book is for

This book is for web developers using C# and .NET as their backend. You are required to have at least a beginner's knowledge of the language. The book will also serve experienced web developers using WebForms, who may consider porting their app to Blazor technology. Knowledge of HTML is essential for this book.

What this book covers

Chapter 1, Introducing Blazor, gRPC, and Source Generators, focuses on providing an introduction to the Blazor WebAssembly framework, gRPC technology, and source generators.

Chapter 2, Creating a Blazor WebAssembly Application, covers Blazor WebAssembly and Razor components. It shows you how to create two-way data binding, pass values deeper to the components, and re-render components when needed.

Chapter 3, Creating a Database Using Entity Framework Core, covers using Entity Framework Core as a database provider and creating code-first database structures. The topic of mapping objects between each other is also explored.

Chapter 4, Connecting Client and Server with REST API, covers creating endpoints in the Blazor application and exposing the REST endpoints for the client part of the application. The topic of generic Razor components and how to consume REST API within them is also covered.

Chapter 5, Building gRPC Services, explores two options for implementing gRPC in .NET projects, in addition to the syntax of a protocol buffer language.

Chapter 6, Diving Deep into Source Generators, covers the topic of creating source generators to generate some code for us with the aim of easing the development process. Parts of code that can or cannot be generated will also be explained.

Chapter 7, Best Practices for C# and gRPC, covers the code-first approach for gRPC and discusses why gRPC can't replace REST API.

To get the most out of this book

You will need the .NET 6 SDK version installed on your computer. The latest version of the SDK is included in the Visual Studio 2022 Community installation. All the code examples have been tested using .NET 6, and Visual Studio 2022 Community and Professional on Windows 11. However, they should work with the future version of .NET and also with Visual Studio Code and Visual Studio for Mac.

Software/hardware covered in the book	Operating system requirements
Visual Studio 2022 (Community)	Windows, macOS*, or Linux*
Microsoft SQL Server 2019	
.NET 6	
Blazor	
Entity Framework Core	
gRPC	

* For Linux users, you can use Visual Studio Code instead of Visual Studio 2022 Community. For macOS users, Visual Studio for Mac is the right tool. The SQL Server can be replaced with an in-memory database.

If you are using the digital version of this book, we advise you to type the code yourself or access the code from the book's GitHub repository (a link is available in the next section). Doing so will help you avoid any potential errors related to the copying and pasting of code.

If you have not worked with Visual Studio before, I recommend looking for some tool overview tutorials on the Microsoft site or YouTube.

Download the example code files

You can download the example code files for this book from GitHub at `https://github.com/PacktPublishing/gRPC-Powered-Blazor-WebAssembly-Development`. If there's an update to the code, it will be updated in the GitHub repository.

We also have other code bundles from our rich catalog of books and videos available at `https://github.com/PacktPublishing/`. Check them out!

Download the color images

We also provide a PDF file that has color images of the screenshots and diagrams used in this book. You can download it here: `https://packt.link/xNJtA`.

Conventions used

There are a number of text conventions used throughout this book.

`Code in text`: Indicates code words in text, database table names, folder names, filenames, file extensions, pathnames, dummy URLs, user input, and Twitter handles. Here is an example: "In the preceding code, `SyntaxReceiver` contains the `OnVisitSyntaxNode` method."

A block of code is set as follows:

```
message Person {
    int32 id = 1;
    string name = 2;
    repeated int32 moviesIds = 3;
}
```

When we wish to draw your attention to a particular part of a code block, the relevant lines or items are set in bold:

```
[ProtoContract]
public interface IPersonRepository
{
    ValueTask<Person> CreateAsync(Person person, CallContext
    contex = default);
}
```

Any command-line input or output is written as follows:

```
Install-Package Microsoft.EntityFrameworkCore
dotnet add package Microsoft.EntityFrameworkCore.Design
```

Bold: Indicates a new term, an important word, or words that you see onscreen. For instance, words in menus or dialog boxes appear in **bold**. Here is an example: "Click the right mouse button on **Dependencies** in the **MediaLibrary.Server** project."

> **Tips or important notes**
> Appear like this.

Get in touch

Feedback from our readers is always welcome.

General feedback: If you have questions about any aspect of this book, email us at customercare@ packtpub.com and mention the book title in the subject of your message.

Errata: Although we have taken every care to ensure the accuracy of our content, mistakes do happen. If you have found a mistake in this book, we would be grateful if you would report this to us. Please visit www.packtpub.com/support/errata and fill in the form.

Piracy: If you come across any illegal copies of our works in any form on the internet, we would be grateful if you would provide us with the location address or website name. Please contact us at copyright@packt.com with a link to the material.

If you are interested in becoming an author: If there is a topic that you have expertise in and you are interested in either writing or contributing to a book, please visit authors.packtpub.com.

Share Your Thoughts

Once you've read *Building Blazor WebAssembly Applications with gRPC*, we'd love to hear your thoughts! Scan the QR code below to go straight to the Amazon review page for this book and share your feedback.

https://packt.link/r/1-804-61055-0

Your review is important to us and the tech community and will help us make sure we're delivering excellent quality content.

1
Introducing Blazor, gRPC, and Source Generators

Let us start our journey in this book by first getting to know our three main technological co-travelers: Blazor, gRPC, and source generators.

The **Blazor** framework is an open source web framework developed by Microsoft. It is a free-to-use, **single-page application** (**SPA**) framework that enables a smooth development process while writing both the server and client parts of the application in .NET and C#. It can be used for dynamic websites but can also generate static websites without the need for expensive website hosting.

On the other hand, **Google Remote Procedure Call** (**gRPC**) is a multi-environment open source **Remote Procedure Call** (**RPC**) framework. It uses **HTTP/2** for connecting services between data centers, but also for connecting mobile applications and browsers to backend services. gRPC includes optional features for authentication, bidirectional streaming, health checks, and more, along with support for cross-platform client and server code generation for many languages.

RPC was first developed in the 1970s and implemented in the 1980s. In 2015, Google built gRPC on top of the existing RPC protocol and it became a new standard supported by many languages.

Last but not the least, we will be working with **source generators**, a new feature for **C#** developers, available from **.NET Standard 2.0**. This feature allows us to programmatically read and analyze our code before its compilation and extend or create new code that is included in the compilation process. We can think of source generators as templating mechanisms that can emit C# source code.

Before we move further, we also need to explain what **WebAssembly** (**WASM**) stands for. WASM is a binary-code format for portable executable programs. It enables creating high-performant programs running inside a web browser without a close connection to the browser itself. This means that WASM can also run in other environments. WASM is an open standard supported by many languages and operating systems.

The **Blazor WebAssembly** framework will act as the glue that binds together all the technologies we will be working with to build our application. In this book, we will get to know each of these technologies better and understand their purpose. We will also build a real, highly performant gRPC-powered Blazor WebAssembly application using all of these technologies.

By the end of this chapter, you will have learned about the Blazor WebAssembly framework and how it can extend your focus from backend development to frontend development, without the need for a different programming language. You will learn about the gRPC protocol and how you can use it instead of **Representational State Transfer** (**REST**) in server-to-server and browser-to-server communication. In the last part of the chapter, you will learn about techniques to avoid writing repetitive code using source generators.

In this chapter, we will cover the following topics:

- Using the Blazor framework to create websites

- Understanding the REST API as the default Blazor communication interface

- What is gRPC and how is it different from REST?

- What are source generators and how do they work?

Technical requirements

All projects in this book will use *Windows 11*. If you are using an older version of Windows, you can expect some differences.

All of the projects are built using **Microsoft Visual Studio 2022 Community Edition** version *17.1.1*. If you need to install Visual Studio 2022, select **Free Download** under **Community** at `https://visualstudio.microsoft.com/downloads/`. Run the installer. When prompted, select the option for **ASP.NET and web development**.

Once the installation process completes, the welcome window of Visual Studio will open. You can click the **Continue without code** link on the right side of the window to open the Visual Studio IDE without any solution.

> **Tip**
> If you already have Visual Studio Professional or Enterprise edition, everything should work the same way. However, the newer version (or preview) of Visual Studio can handle some code actions differently. Rider from JetBrains can be used as well.

Our projects require **.NET 6.0**. If you have installed Visual Studio 2022, .NET 6.0 should be already installed. You can check the installed version by running the following command in **Windows PowerShell**:

```
dotnet --version
```

If the required version is not installed, download the installer from `https://dotnet.microsoft.com/en-us/download/dotnet/6.0`. After running the installer, you can open Windows PowerShell again and run the preceding command. The command should print the correct version of .NET installed on your computer.

Our project in this book will use **SQL Server 2019 Developer Edition** for the database. Microsoft provides a free version of SQL Server for developers. To install SQL Server Developer, open `https://www.microsoft.com/en-us/sql-server/sql-server-downloads` and download the **Developer** edition. Run the installer and when prompted, select **Basic installation**. Accept the Microsoft SQL Server License Terms. Click the **Install** button to let the installer complete the installation. You can customize the installation by clicking on **Install SMSS**. **SQL Server Management Studio** (**SSMS**) is a tool to manage SQL databases and is highly recommended to install.

> Tip
> Any edition of the SQL Server can be used to run this project. However, versions older than 2015 are not recommended.

In this book, we will use Blazor WebAssembly in .NET 6 for all projects.

You can find the complete source code for this book on GitHub at `https://github.com/PacktPublishing/gRPC-Powered-Blazor-WebAssembly-Development`.

All the code for this chapter can be found at `https://github.com/PacktPublishing/gRPC-Powered-Blazor-WebAssembly-Development/tree/main/ch1`.

Using the Blazor framework to create websites

The Blazor framework can be used in different hosting models. **Blazor Server**, WebAssembly, and Hybrid hosting models are supported, and each of them has different use cases and advantages and disadvantages.

The Server hosting model uses **SignalR** technology to send data from the client to the server. Then the server does the work and sends back the data required to update the UI of the application.

The WebAssembly hosting model has its client part of the application downloaded to the client with all the binaries needed to run the application. The WebAssembly application then does all the work on the client's computer. This approach is faster but requires downloading larger files to the client. Since the end of 2017, all major browsers support WebAssembly, including mobile browsers. For old browsers, WebAssembly needs to be compiled to `asm.js` by JavaScript *polyfill*.

The **Blazor Hybrid** model can also be used to blend the desktop and native mobile frameworks. The components in the Hybrid model have access to mobile phone capabilities.

In this book, we will be using the Blazor WebAssembly hosting model. We are not focusing on mobile app development, so the Hybrid model is not useful for our purposes. Also, we want to create an application in which we don't care as much about every millisecond of loading time, as we care about the speed of the application. And because WASM does its work inside the browser, all the re-rendering and UI updates are faster than in the server hosting model.

Using Blazor WebAssembly for single-page applications

Since Blazor is a dynamic framework; it can be used for an SPA. Blazor re-renders only the parts of the page that need to be changed, without the need to request all HTML for the page from the server and reload the page in the browser. This behavior creates a much better user experience on the website.

During the rendering process of the page, the browser creates the **Document Object Model (DOM)**. The model is a render tree graph of each element used on the page. At the same time, Blazor creates its own **virtual DOM**. This virtual DOM holds more information than the DOM of the browser. There is a place for each element that can possibly be rendered on the page if the conditions change. When the user requests some change on the page, such as clicking on a link or submitting a form, Blazor re-renders the virtual DOM. These changes are then compared to the browser DOM. Only the difference is rendered to the client.

Re-rendering non-change elements costs CPU time and memory and slows down the overall process on the client (browser). However, comparing DOM models works much faster. Most modern SPA frameworks use their own kind of virtual DOM to prevent unnecessary updates in the browser.

Harnessing advantages of progressive web applications using Blazor

One of the other benefits of using Blazor for creating a website is native support for **progressive web applications (PWAs)**. Such applications can be packed to run in the WebView on Windows and mobile phones (iOS and Android) and look like native apps. Many companies use PWAs as they are easier and cheaper to develop. PWAs can be used to build just one website that renders well across many devices.

Like everything, PWA has disadvantages. Missing the browser's API to communicate with device hardware can lead to requiring a native application. iOS also prevents push notifications. This will, however, likely change in the future.

Understanding how Blazor uses Razor syntax

Razor syntax is a combination of HTML markup and C# code. Using both in one file helps to create a better and faster development process where you can render elements if the code meets some conditions, or easily render tables using `for` loops.

Razor syntax is not new in the .NET world. The syntax is used in Razor Pages and the **model view controller (MVC)**. The main difference is that Blazor uses Razor syntax to create **Razor components**, while MVC and Razor Pages use it to render the whole page. You may find it easy to figure out the difference by looking at file extensions. Razor components use the `.razor` extension, while MVC uses `.cshtml`.

Razor components are the main building blocks for the Blazor application. Each component creates its own reusable **user interface** (**UI**) that can be used anywhere on the website. A single component can consist of one `.razor` file (such as MyComponent.razor), which can include both HTML markup and C# code with some logic. Optionally, the logic of the component can be extracted to a partial class with the same name as the Razor component. The convention is to name the file the same as the component with a `.cs` extension, such as MyComponent.razor.cs:

MyComponent.razor

```
<h1>Here is my @Title</h1>
```

MyComponent.razor.cs

```
@code {
    [Parameter] public string Title { get; set; }
}
```

Razor components can be also packed with custom **Cascading Style Sheets** (**CSS**) files to provide some specific component design. As the component author, you can create custom CSS for your component that will not interfere with the other components, but still allow the developer to modify styles with the global CSS. This ability is a huge advantage over MVC or other server-side frameworks. Another advantage is that the CSS included in the component is downloaded to the client only when the component is used. This advantage can lead to lowering the amount of data needed by the client.

> **Interesting fact**
> The name of the Blazor framework was created by combining the words *Browser* and *Razor*.

Now that we have some understanding of how Blazor can be used to create fast and responsive websites with reusable UI components written in Razor syntax, let us learn about how JavaScript can be used in Blazor web applications.

Using JavaScript in Blazor

Most modern web applications are using **JavaScript** frameworks, such as Vue.js, Angular, React, and others. Blazor is a framework in itself. However, this does not mean that we can't use JavaScript API in our Blazor applications. On top of that, we can use JavaScript to call our C# code.

The process of these calls is called **JavaScript interoperability** (**JS interop**).

Calling JavaScript APIs

All JS interop calls are by default *asynchronous*. On Blazor Server, the JS interop calls are sent over the network. Regardless of the called code, the call is done asynchronously. However, there are synchronous JS interop calls for the Blazor WebAssembly hosting model.

Calling .NET from JavaScript

To allow calling the .NET method from JavaScript, the method must have a `JSInvokable` attribute. This attribute creates a binding between JavaScript calls and .NET. It allows us to specify a name for JavaScript calls, which is different from the original method name.

Calling the .NET method from JavaScript is then done by calling `DotNet.invokeMethodAsync` returning a JS `Promise` or `DotNet.invokeMethod` returning the result of the operation. The synchronous version (`invokeMethod`) does not support the Blazor Server hosting model.

With knowledge about interactions between the Blazor application and JavaScript APIs, you can create rich, dynamic websites that have access to all JavaScript features in the browser and can also benefit from the C# language and capability. Let us now learn a bit about the REST API, which is the default Blazor interface for communication between the browser (client) and the server part of the application before talking about gRPC. Understanding REST API will provide us with a better context to understand the workings and benefits of gRPC.

Understanding the REST API as the default Blazor communication interface

REST is an architecture for distributed systems. It is an easy way to read, edit, or delete data on the server using HTTP calls from the client. Distributed architecture in this context means that multiple servers can handle different parts of the application and communicate through the network. The communication can occur between the browser and the server, but also between multiple servers, where each of them can provide different functionality, such as logging, caching, authentication, and so on.

A REST API (also called a RESTful API) uses URL calls to communicate and is stateless. Thus, every call to the resource (URL) has to have all information to be completed. Each call can be marked as cached, which can lead to performance improvement between the client and the server.

For data manipulation, REST specifies four basic operations: **Create, Read, Update, and Delete (CRUD)**. Each operation has its own method (`POST`, `GET`, `PUT`, and `DELETE`). The basic data structure used to transfer data between the client and server is JSON or XML file format, or multipart form data in the case of submitting HTML forms or files.

Examples with REST API calls can be found in the GitHub repository for this chapter, the link to which has been provided in the *Technical requirements* section.

Blazor uses the REST API as the default interface for communication between client and server. The client part of the application has a defined `HttpClient` with a base URL address to the server. Blazor uses this client to call the server to get data to the client, or to store submitted data back on the server.

Now that we have an understanding of the REST API and how Blazor uses it to transfer data between the client and server, let us learn how we can improve the speed and performance of the communication using the gRPC protocol.

Understanding gRPC and how it is different from REST

gRPC is a way to communicate between services. You can think about it as a kind of API that sends the request to the server and expects some data to be returned. The main difference between a REST API and gRPC is the setup and the way it handles and transports data.

gRPC enables **contract-based communication** between the client and server. You can imagine this communication to be like a handshake between the client and server over available methods that can be used in future calls. Unlike in a REST API, in gRPC, the client and server share the same configuration, so both ends of the communication have the exact knowledge of the data structure of the transferred objects. The contracts in gRPC are called **protocol buffers**.

Another big difference between REST and gRPC occurs in terms of communication type. As we learned in the previous section, the REST API is used for communication with JSON or XML data serialization. gRPC, on the other hand, uses a binary stream, which is much more efficient. One advantage of the binary stream is the compression of the symbols. So, the same message is transferred using fewer bytes. Because of the contract, one side of the communication knows the data structure sent by the other side. There is no need to send the structure with the data.

In the following code, you can see the definition of the `CarManufacturer` class in C#:

```csharp
public class CarManufacturer
{
    public string Name { get; set; }
    public double YearSales { get; set; }
}
```

The preceding code will be translated to the following JSON format when sending a list of data over the REST API:

```json
[
    {
        "Name":"Ford",
        "YearSales":155000
    },
```

```
{
    "Name":"Audi",
    "YearSales":264820
},
{
    "Name":"BMW",
    "YearSales":195700
}
]
```

In the preceding code of the JSON data, we can see the repetitiveness of the object property names. Because of this repetition, the size of the data transferred over the network increases significantly with each item we want to transfer.

To solve this, the contract in gRPC determines which property goes first, second, and so on. We do not need to send this information over the network, as the client and server know the property names already. The binary stream would then contain only the values that we need to translate, which results in a huge saving of bytes. We make this process possible through contracts or protocol buffers.

How protocol buffers work

Protocol buffers comprise a language and platform-neutral mechanism for serializing structured data. They also generate native language bindings to many programming languages. Protocol buffer files have a .proto extension and contain a definition language that defines the object and methods generated on the interface for client and server. An example of the type defined in the proto file can be as follows:

```
message DotNetVersion {
    int32 id = 1;
    optional string name = 2;
    double version = 3;
}
```

The preceding code will generate a C# class with the same name and properties:

```
public class DotNetVersion
{
    public int Id { get; set; }
    public string Name { get; set; }
    public double Version { get; set; }
}
```

gRPC works by defining services, objects, and methods that can be called. Supported languages provide the mechanism for generating clients and servers from specified .proto files. On the client side, a client is generated providing all methods defined in the file. The server then generates the same methods.

C# implementation of gRPC generates the server as an abstract partial class with virtual methods. Each of these methods needs specific implementation to do the required job. The following is an example of the .proto file using our defined DotNetVersion message from the previous example:

```
service DotNetVersionService {
rpc GetVersion(DotNetVersionRequest) returns (DotNetVersion) {}
}
message DotNetVersionRequest {
    int id = 1;
}
```

In the preceding code, we are defining a service class called DotNetVersionService with one method called GetVersion.

gRPC supports multiple languages, such as C#, C++, Dart, Go, Java, Kotlin, Node, Objective-C, PHP, Python, and Ruby. Each language has its own implementation in terms of how it generates the client and server from protocol buffers. However, the generated clients and servers have the same methods, so they are able to communicate with each other. It does not matter whether your server is in C#. You can have clients in C#, Ruby, and PHP together, and they will work with gRPC because gRPC is **platform agnostic**.

The following diagram shows communication between the server written in C# and two clients, one in PHP and the second in Ruby:

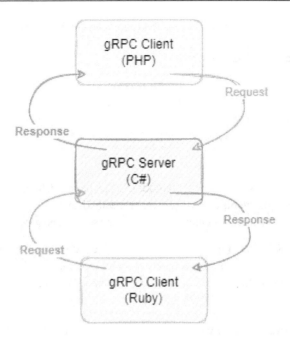

Figure 1.1 – gRPC server and client communication

In the preceding diagram, the clients can be other servers or different types of clients such as browsers, mobile apps, or other smart devices. If gRPC can send the same data in a smaller size and has better-typed clients and servers, why are we still using REST for the web?

Why we still use REST

The problem sits in the HTTP/2 requirement for gRPC communication and basic design. gRPC was primarily designed to communicate in a server-to-server environment, involving types, objects, methods, and classes. Although modern JavaScript can have almost all of this, the support for the HTTP/2 protocol is not fully available in every browser.

gRPC support in .NET and C# began in 2015 with the `Grpc.Core` package. At that time, the implementation was based on the gRPC C Core native library. In May 2021, the team introduced a new implementation based on C# with native support for HTTP/2 protocol called `grpc-dotnet` (`Grpc.Net.Client` and `Grpc.AspNetCore.Server` NuGet packages). The older implementation runs in maintenance mode and will become obsolete in May 2023 as per information available at the time of writing this book (`https://grpc.io/blog/grpc-csharp-future/`).

By now, you know why gRPC is sometimes preferable over REST, especially through the use of protocol buffers, but not always. We will discuss these subjects in more detail in *Chapter 5, Building gRPC Services*, and *Chapter 7, Best Practices for C# and gRPC*. In the next section, we will focus on a new Microsoft technology, **source generators**, which will help us a great deal to create our Blazor WebAssembly application with gRPC.

What are source generators and how do they work?

Can you imagine code that writes code for you? With new features started in C# 9, such code is possible. Let's dive into source generators, which can do exactly what the name says: generate source code.

Source generators are part of the **.NET Compiler Platform (Roslyn) SDK**, which makes them available everywhere dotnet code can be developed, no matter what IDE you use. The generators read code that you write and can generate some additional code that will be added to the compilation and emitted in the resulting .dll.

First, let me say what source generators *can't* do. They can't modify existing code. It is in the design of source generators to be able to add additional files to the compilation, but they can't modify or remove existing files. Whether this is good or bad is a long-running discussion.

What source generators *can* do is they can read a **compilation object** representing all code that is being compiled. Then, they get the compilation object syntax tree and semantic models (similar to analyzers). Knowing what code is compiled, the source generators can emit other code generated as a string. This code is then injected back into the compilation process. We can get a better idea of how this works with the following simple diagram:

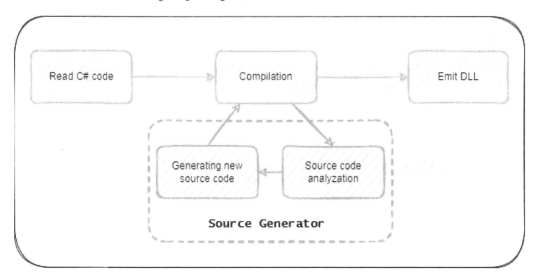

Figure 1.2 – Source generator schema

The source generator steps into the compilation process and analyzes all source code written by the developer that is in the compilation. After the analysis, new code is generated as a string and injected back into the compilation. This process happens only one time during the compilation – it is not recursive. This means that generators can't analyze code generated by other generators. Generators also can't see other generators. Only the code written by the developer – the code that was first in the compilation – is inspected by the generators.

We can write as many generators as we want. However, we should also write some other code, to have a functional application. Once the compilation begins, all of the written source code is read. Then, each of the generators is called once to generate additional source codes. The calls to the generators are parallel. When all generators finish their tasks, the result of the generation is packed with the original source code and then emitted to the **dynamic link library** (**DLL**).

In this section, we have learned about source generators and how they work with our code. To understand more about how source generators read files and how deep their analysis is, we will learn about syntax trees and semantic models, and how to use all this in a real application, in *Chapter 6, Diving Deep into Source Generators*.

Summary

After reading this chapter, you should know about the Blazor framework and the benefits of using this framework for C# developers over other SPA JavaScript frameworks, because you can use the same language for both frontend and backend development.

We also covered how gRPC is better than, and can replace, the REST API in terms of the growing popularity and platform neutrality, and also lower network usage. You should also have some idea about how source generators can ease development.

Thus, you are now ready to start developing Blazor WebAssembly SPA applications and inspecting the code that goes into the compilation process. You can read `.proto` files and use them for gRPC communication, for better performance compared to REST.

In the next chapter, we will cover Razor components and syntax, look at page routing, and introduce an overview of our demo project.

Further reading

If you want to go deeper into some of the topics of this chapter, the following resources can provide more information:

* For more information on *Blazor*, refer to `https://www.blazor.net`.
* For more information on *Blazor Hybrid*, refer to `https://docs.microsoft.com/en-us/aspnet/core/blazor/hybrid/`.
* For more information on *.NET*, refer to `https://dotnet.microsoft.com`.
* For more information on *SPA*, refer to `https://en.wikipedia.org/wiki/Single-page_application`.
* For more information on *PWA*, refer to `https://web.dev/progressive-web-apps/`.
* For more information on *WebAssembly*, refer to `https://webassembly.org/`.

- For more information on *gRPC*, refer to `https://grpc.io/`.

- For more information on *Source Generators*, refer to `https://devblogs.microsoft.com/dotnet/introducing-c-source-generators/` or `https://docs.microsoft.com/en-us/dotnet/csharp/roslyn-sdk/source-generators-overview`.

2
Creating a Blazor WebAssembly Application

Blazor WebAssembly applications use Razor components as building blocks. The component itself is a reusable part of the UI that can be used anywhere in our application. Razor components are a combination of *HTML* and *C#* code; however, the HTML part of the component is compiled into C# code and emitted in the final DLL. The component files are not standalone files like in standard `.html` pages.

In this chapter, we will learn about Razor components. We will understand how these components are built, how they communicate with other components on the page, and what their life cycle is. We will learn how to create and use components, how to add parameters to components, and how these components can notify parents that something has happened. This is useful when we want to notify the parent about the data changing, a button or link being clicked, and so on.

We will also learn two ways of defining C# code for the components, in addition to learning how to specify the route URL for the component and how to read data from the website URL address.

Finally, we will take a look at the project we want to build and start building it using the default Blazor WebAssembly App project template provided by Microsoft.

By the end of this chapter, you will understand how to create a new project for a Blazor WebAssembly application using Visual Studio. You will also understand the concept of Razor components, how they communicate, and how Blazor uses these components to render the content on the websites.

In this chapter, we will cover the following topics:

- Learning how to write Razor syntax
- Creating Razor components
- Understanding page routing in Blazor
- Project overview and preparation

Technical requirements

All the code for this chapter can be found at `https://github.com/PacktPublishing/gRPC-Powered-Blazor-WebAssembly-Development/tree/main/ch2`.

Learning how to write Razor syntax

Razor syntax is a combination of HTML, Razor markup, and C#. The syntax is similar to other JavaScript SPA frameworks, such as Vue.js, React, and Angular. The HTML in the Razor component is rendered the same way as it is in the HTML file. In the Razor component, you can fluently move between HTML markup and C# code.

Razor uses the @ symbol to determine the Razor syntax. If the @ symbol is followed by a reserved keyword, then the code is transitioned to Razor syntax. This Razor syntax can add dynamic logic to our components. The expressions can be implicit, explicit, inline, or in the form of code blocks.

Writing implicit Razor expressions

Implicit Razor expressions start with the @ symbol, followed by C# code:

```
<span>It is @DateTime.Now</span>
```

In the preceding code, @ is followed by the C# code directly. Implicit Razor expressions can't contain spaces or generic calls, because < and > symbols in the generic call are translated as HTML in the Razor syntax. The only exception is the `await` keyword, which can be followed by a space. If the method is used in the implicit Razor syntax call, the spaces can be used in the arguments.

Writing explicit Razor expressions

Explicit Razor expressions start with @, followed by parentheses:

```
<span>It was @(DateTime.Now - TimeSpan.FromDays(7))</span>
```

The preceding code renders the date in the last week. Explicit Razor expressions are used when we can't access the result value directly in the property or field or when we need to calculate some result, use generic calls, or prevent the value in the HTML string from being concatenated. Here is an example:

```
@{
var addr = new Address(Street: "Fairmont St NW", City:
"Washington", "PostalCode": "20009");
}
<span>The address is@addr.Street @addr.City @addr.PostalCode</span>
```

The preceding code will render the following output HTML:

```
<span>The address is@addr.Street Washington 20009</span>
```

In the preceding output, `addr.Street` is not replaced by the property value because the term may comprise the syntax of an email address – that is, `is@addr.Street`. When we need to use a directive just after a letter, we need to use the explicit Razor syntax:

```
<span>The address is@(addr.Street) @addr.City @addr.
PostalCode</span>
```

As expected, the preceding code will generate the output HTML:

```
<span>The address isFairmont St NW Washington 20009</span>
```

> **Important information**
>
> If the @ symbol follows the letter, it is not translated to the Razor directive. If you need to render @ in the HTML, you can escape the symbol with another @.
>
> For example, @@MyUserName will render in the HTML as @MyUserName.

Writing inline expressions

Inline expressions are all of the expressions on a single line. Both implicit and explicit expressions are inline expressions.

Writing code block expressions

When creating our Razor components, we need to use some functionality other than just the properties or fields from C#. The code block expressions start with the @ symbol and are enclosed within curly brackets:

```
@{
var title = "About Blazor WebAssembly";
}


<h2>@title</h2>


@{
title = "Razor syntax";
}
```

```
<h2>@title</h2>
```

The preceding code will generate the following output HTML:

```
<h2>About Blazor WebAssembly</h2>
<h2>Razor syntax</h2>
```

As you can see in the preceding output, the generated HTML consists of two headlines.

Block expressions can also be used to create local functions to render repeated HTML code:

```
@{
void RenderTitle(string title)
{
<h2>@title</h2>
}
RenderTitle("About Blazor WebAssembly");
RenderTitle("Razor syntax");
}
```

The output HTML generated from the preceding code will be the same as in the first example in this section. We do not need to modify the variable each time we want to render the title because we can use the RenderTitle function defined in the code block. This example shows simple code, but in the real world, the local function will contain more complex code.

Writing control structures

Control structures are specific types of code blocks. They are used when the code block has some specific meaning to the code, such as conditions, loops, and so on.

Conditions

We can use the if directive to start a code block with conditions:

```
@if (chapterNumber == 1)
{
  <p>This is the first chapter.</p>
}
else if (chapterNumber == 6)
{
  <p>This is the last chapter.</p>
```

```
}
else
{
    <p>Chapter @chapterNumber</p>
}
```

The preceding code shows a full `if - else if - else` statement for rendering paragraph information for each chapter in this book. The same condition can be applied by the `switch` statement:

```
@switch (chapterNumber)
{
    case 1:
        <p>This is the first chapter.</p>
        break;
    case 6:
        <p>This is the last chapter.</p>
        break;
    default:
        <p>Chapter @chapterNumber</p>
        break;
}
```

The preceding `switch` statement will generate the same HTML result as the `if` statement.

Loops

The following loop types are supported in Razor syntax:

- `for`
- `foreach`
- `while`
- `do while`

Each loop starts with the @ symbol. Looping through a collection of items helps you in situations where you want to apply some repetitive code to each item in the collection. The following example shows the looping of `chapters`:

```
@{
    var chapters = new Chapter[]
    {
```

```
    new Chapter(1, "Learning Razor syntax"),
    new Chapter(2, "Creating Razor components"),
    new Chapter(3, "Understanding page routing in Blazor"),
    ...
  };
}
```

The preceding code defines a new `chapters` variable that will be used in the following loop examples:

- The following is an example of a `for` loop:

```
@for (var i = 0; i < chapters.Length; i++)
{
    var chapter = chapters[i];
    <h2>Chapter no. @chapter.Number: @chapter.Title</h2>
}
```

- The following is an example of a `foreach` loop:

```
@foreach (var chapter in chapters)
{
    <h2>Chapter no. @chapter.Number: @chapter.Title</h2>
}
```

- The following is an example of a `while` loop:

```
@{ var i = 0; }
@while (i < chapters.Length)
{
    var chapter = chapters[i];
    <h2>Chapter no. @chapter.Number: @chapter.Title</h2>
    i++;
}
```

- The following is an example of a `do while` loop:

```
@{ var i = 0; }
@do
{
```

```
    var chapter = chapters[i];
    <h2>Chapter no. @chapter.Number: @chapter.Title</h2>
    i++;
}
while (i < chapters.Length)
```

As you can see, the Razor syntax is similar to the C# syntax. The only difference is in the transition between C# code and HTML code.

Razor syntax allows some other keywords to be used in the code blocks, such as @using (for disposable objects), @try, catch, finally, and @lock. The HTML and C# comments are also allowed.

Writing top-level directives

Razor **directives**, which are at the top of the file, control many aspects of the component. They can change the way the component is compiled, how the component is used on the website, or who can view the component's output. The directives should be at the top of the file, but the order of the directives is not defined. Here is the list of these top-level directives:

- @page: This directive specifies the page where the component is rendered. We will learn more about this in the following sections. Here is an example of the directive for the home page:

  ```
  @page "/home"
  ```

- @namespace: The @namespace directive can override the namespace for the component. The default component namespace is based on the folder structure. Here is an example of namespace overriding:

  ```
  @namespace DemoProject.MyLibrary.Components
  ```

- @inherits: When we want to change the base class of the component, we need to use the @inherits directive, as shown in the following example:

  ```
  @inherits CustomComponentBase
  ```

- @implements: Our component can implement any interface. This directive can be used multiple times. Here is an example of implementing two interfaces:

  ```
  @implements IDisposable
  @implements ISortable
  ```

- @layout: The @layout directive is only used on the component that contains the @page directive. The default layout is used if nothing has been set. If the page directive is missing, the @layout directive is ignored. Here is an example of the layout directive setting the layout to LoginLayout:

  ```
  @layout LoginLayout
  ```

- @attribute: The @attribute directive adds a class-level attribute to the component class. This directive can be used multiple times. Here is an example of marking a component for only authorized viewing:

  ```
  @attribute [Authorize]
  ```

- @using: This directive can be used multiple times to import namespaces to the component scope:

  ```
  @using DemoProject.Model
  ```

- @typeparam: The @typeparam directive is mostly used in combination with the generic base class in the @inherits directive. If the base class contains multiple generic types, the order of the @typeparam directive must reflect the order of the generic types:

  ```
  @implements MyBaseClass<TItemModel, TFilterModel>
  @typeparam MyItemModel
  @typeparam MyFilterModel
  ```

- @inject: The @inject directive injects service from the dependency injection container into the component:

  ```
  @inject HttpClient Client
  @inject IJSRuntime JS
  ```

  The preceding code shows how to inject an HttpClient into the Client property and IJSRuntime into the JS property.

Writing inline directives

Inline directives are used in the HTML or Razor components. These directives include the following:

- @attributes: These directives represent a collection (Dictionary<string, object>) of HTML attributes that we want to render in the child component that has not specified these HTML attributes as component parameter properties:

  ```
  @{
      Dictionary<string, object> AdditionalAttributes =
          new Dictionary<string, object()
  ```

```
    {
        { "tooltip", "Set your full name." },
        { "required", "true" },
        { "data-id", 150 }
    };
    }
    <input @attributes="AdditionalAttributes" />
```

- @bind: This directive creates two-way data binding for a component with user input:

```
@{ public string UserName { get; set; } }
<input type="text" @bind-Value="UserName" />
```

- @on{event}: This directive adds an event handler for the event specified (click, change, and so on):

```
<button @onclick="OnClickHandler">Click here!</button>
```

- @key: This directive specifies the unique key used to render the collection of data in loops (for, foreach, while, and do while):

```
@for (var i = 1; i <= 10; i++)
{
    <p @key="i">This is row @i</p>
}
```

Such keys are used when updating the rendered data to update a correct item of the loop, instead of updating the whole loop.

- @ref: This directive captures a reference to the component or HTML element:

```
<ConfirmBox @ref="myConfirmBox" />
@code {

    ...

    myConfirmBox.OnConfirm(...);

    ...

}
```

@ref can be used to trigger JavaScript events.

Now that we know how to write Razor syntax, let's learn how to use Razor syntax and create fully functional Razor components.

Creating Razor components

Razor components are the basic building blocks of our Blazor WebAssembly application. Each Razor component is represented by a `class` with a name similar to the filename. In the component itself, we can use C#, HTML, and Razor markup. This class is generated automatically for us, but we can create it ourselves. The component is downloaded to the browser as a part of the DLL for the whole application.

Razor components can be anything, from simple standard HTML for a headline, to more complex elements, such as tables that contain data, forms, or anything that can be rendered on the website. Each component can use other components.

> **Note**
>
> A Razor component name must be in Pascal case. Naming a component as `pageTitle` is not a valid name for the Razor component, because the p part is not a capital letter. `PageTitle` will be a valid name. The file extension for Razor components is `.razor`.
>
> *To prevent errors, you should also name components differently from existing HTML tags.*

A simple component representing the page title can be created with just a single line of HTML code in the `Headline.razor` file:

Headline.razor

```
<h1>Page title</h1>
```

The preceding code shows the simplest type of Razor component. In the background, `public partial class Headline` is created for us.

Using components in other components

The `Headline` component itself can show just a small amount of data. We need to use this `Headline` component in other components. We can do this with standard HTML markup. The class name of the component is also the name of the HTML element. The following code shows how to use our `Headline` component in different components:

PageHeader.razor

```
<div class="header">
<Headline />
</div>
```

The preceding code will create another Razor component using our `Headline` component as part of the rendered HTML.

The Razor component class is created in the namespace to reflect the current folder structure of the project. If the `Headline` component exists in a different folder than the `PageHeader` component, the usage of the component must reflect the namespace of the `Headline` component:

```
<Structures.Headline />
```

In the preceding example, the `Headline` component is in the `Structures` folder. The `Structures` folder is in the same folder as the `PageHeader` component.

To prevent specifying the component namespace in the markup, we can add the `@using` directive to the parent component, or create a file called `_Imports.razor` and add `@using` directives there. The following code shows how to make the `Headline` component available for all other components in the same project:

_Imports.razor

```
@using SampleApp.Components.Structures
```

This simple component can be used for static HTML markup, such as loaders and icons.

> **Note**
> `@using` directives in C# files (`.cs`) are not applied to Razor components (`.razor`) and vice versa. The `_Imports.razor` file's `@using` directives are only applied to Razor components.

When we need to render some dynamic data, we need to load it inside the component itself or pass it to the component from the parent component.

Passing parameters to components

Razor components use parameters to create dynamic content and conditional rendering. Parameters are public properties of the component class. These parameters are decorated with either the `Parameter` attribute or the `CascadingParameter` attribute. Parameters can be any type of object in C#, such as simple types, complex types, functions, and so on.

There are two types with special meaning for Razor components:

- Event callback
- `RenderFragment`

We will discuss these two types in the *Creating Razor components* section.

Let's modify our `Headline.razor` component from a simple component with static text to a more dynamic component that can show data from parameters:

```
<h1>Page @Title</h1>
@code {
[Parameter]
public string Title { get; set; }
}
```

To use the preceding component in our `PageTitle` component, we will need to modify the line with component initialization using the following syntax:

```
<Headline Title="About us" />
```

The preceding code shows a parameter called `Title` that's been added to our component. This parameter can be used to pass any string value from the parent component. The value is then used in the HTML markup. The `Headline` component will render **Page About us** text on the screen.

> **Note**
>
> At the time of writing this book, Razor syntax does not support the required attributes on the component parameters. In the latest version of Visual Studio, the `EditorRequired` attribute was introduced to help developers with this problem. The `EditorRequired` attribute tells the IDE that the marked parameter should be filled in when using the component. The **RZ2012** warning is generated, but it will not prevent building the application. The following warning will be generated if we mark the `Title` parameter of our `Headline` component with the `EditorRequired` attribute and don't fill in the parameter: *Component 'Headline' expects a value for the parameter 'Title', but a value may not have been provided.*

The component can have more parameters and can also read parameter values from its route (URL address) or query values. We will learn more about this in the *Understanding page routing in Blazor* section. The component should not update any of the `Parameter` or `CascadingParameter` properties by itself. To update the parameter, the event callback should be triggered.

In the next section, we will look at the `RenderFragment` parameter before covering event callbacks.

Creating components with child content

A common scenario when creating Razor components is the need to pass some additional content from the parent component. This content can be HTML inside the `button` tag, a `figure` element to wrap the image with its description, or a more complex HTML structure.

We can tell our component to render child content by providing a parameter whose type is RenderFragment and whose name is ChildContent. The following example renders complex HTML with content from the parent component:

Product.razor

```
<div class="product">
@ChildContent
</div>
@code {
[Parameter]
public RenderFragment ChildContent { get; set; }
}
```

The preceding code shows a Product component, which is a simple HTML div element with a class attribute and ChildContent parameter. This component can be used in other components to render a unified structure:

List.razor

```
...
<Product>
<div class="image">
<img src="demo_image.jpg" alt="Demo image" />
</div>
<div class="name">Product name</div>
<div class="info">Description & Price</div>
</Product>
...
```

The preceding code creates the List component using the Product component. The List component can contain much more code and render the Product component in the loop.

In this example, we can see that we do not need to specify the ChildContent attribute when using the Product component. Razor will take any content between the opening and closing tags of the component and pass it as a ChildContent parameter automatically.

In the List component, we need to write all the HTML that we want to render inside and repeat that for each product we want to render. This can lead to potential mistakes when reusing the component as it is easy to mistype any class name.

To help solve this, we can define as many `RenderFragment` parameters as we want. However, these parameters must be named by the caller component. In the following example, we are modifying the code of the `Product` component to allow us to use three sections – `Image`, `Name`, and `Info`:

Product.razor

```
<div class="product">
<div class="image">@Image</div>
<div class="name">@Name</div>
<div class="info">@Info</div>
</div>
@code {
[Parameter]
public RenderFragment Image { get; set; }

[Parameter]
public RenderFragment Name { get; set; }

[Parameter]
public RenderFragment Info { get; set; }
}
```

Now that we have defined three parameters with the `RenderFragment` type, we can use them from the parent component by using the parameter name as an HTML tag nested in our component:

List.razor

```
...
<Product>
<Image>
<img src="demo_image.jpg" alt="Demo image" />
</Image>
<Name>Product name</Name>
<Info>Description & Price</Info>
</Product>
...
```

The preceding example is using all of the parameters from the Product component. You can omit any of them and use only the ones you need. The Product component can have conditions to render only the RenderFragment parameters, which are not null.

> **Interesting fact**
>
> The component can have both named RenderFragment parameters and ChildContent parameters. If the parameter is not specified, ChildContent is used. Using at least one of the named RenderFragment parameters leads to the need to name the ChildContent parameter as well.

Now that we know how to pass parameters downstream to child components, let's learn how to update data in the parent component. This is where event callbacks will serve their purpose.

Communicating with the parent component

There are a lot of situations where we need to notify the parent component about some change. These changes can include events such as a click on the element, a change in the element state, double-clicking the element, and dragging and dropping the element. For that purpose, there is an EventCallback or EventCallback<T> property, which can be defined on our component to allow the parent component to react to the events.

The event callback can be without arguments, or with an argument type defined by the T type parameter. The following is a list of predefined argument classes with supported DOM events:

Argument Class	DOM Event
ClipboardEventArgs	oncut, oncopy, onpaste
DragEventArgs	ondrag, ondragstart, ondragenter, ondragleave, ondragover, ondrop, ondragend
ErrorEventArgs	onerror

EventArgs	This argument class is supported by most of the events and holds basic event data.
	General
	`onactivate, onbeforeactivate, onbeforedeactivate, ondeactivate, onfullscreenchange, onfullscreenerror, onloadeddata, onloadedmetadata, onpointerlockchange, onpointerlockerror, onreadystatechange, onscroll`
	Clipboard
	`onbeforecut, onbeforecopy, onbeforepaste`
	Input
	`oninvalid, onreset, onselect, onselectionchange, onselectstart, onsubmit`
	Media
	`oncanplay, oncanplaythrough, oncuechange, ondurationchange, onemptied, onended, onpause, onplay, onplaying, onratechange, onseeked, onseeking, onstalled, onstop, onsuspend, ontimeupdate, ontoggle, onvolumechange, onwaiting`
FocusEventArgs	`onfocus, onblur, onfocusin, onfocusout`
ChangeEventArgs	`onchange, oninput`
KeyboardEventArgs	`onkeydown, onkeypress, onkeyup`
MouseEventArgs	`onclick, oncontextmenu, ondblclick, onmousedown, onmouseup, onmouseover, onmousemove, onmouseout`
PointerEventArgs	`onpointerdown, onpointerup, onpointercancel, onpointermove, onpointerover, onpointerout, onpointerenter, onpointerleave, ongotpointercapture, onlostpointercapture`
WheelEventArgs	`onwheel, onmousewheel`

ProgressEventArgs	onabort, onload, onloaded, onloadstart, onprogress, ontimeout
TouchEventArgs	ontouchstart, ontouchend, ontouchmove, ontouchenter, ontouchleave, ontouchcancel

Table 2.1 – Event callback predefined argument classes

Standard HTML elements in Blazor have defined these arguments for events. The button's onclick attribute will provide the event callback with the MouseEventArgs argument.

You can also define custom event arguments that allow you to transfer any kind of data to the parent component.

The event callback parameter on the child component is attached to the method with the same arguments on the parent component. The following code shows two components that communicate with each other using event callbacks:

LikeButton.razor

```
<button @onclick="@OnClickHandler">
Click to change color to black
</button>
@code {
[Parameter]
public EvenCallback<string> OnClickEvent { get; set; }

private async Task OnClickHandler()
{
   await OnClickEvent.InvokeAsync("black")
}
}
```

The preceding code creates a simple HTML button element with the @onclick directive. Clicking on this button will trigger the OnClickHandler method, notify the parent component about the event, and pass the arguments to the parent component. The parent component can perform some action if this notification occurs:

Parent.razor

```
<Header Title="Color selector" />
<LikeButton OnClickEvent="@OnClickAction" />
```

```
<p>Selected color is @color</p>
@code {
private string color = "white";
private void OnClickAction(string args)
{
   color = args;
}
}
```

In the preceding code, the parent component has attached a custom method for the OnClickEvent parameter. The attached method must consume the same type of parameters as the OnClickEvent type argument.

If you use predefined arguments on supported events, you can call EventCallback<T> directly from the @on... event:

```
<button @onclick="@OnClickEvent">Click Me</button>
@code {
   [Parameter]
   public EvenCallback<MouseEventArgs> OnClickEvent
     { get; set; }
}
```

In the preceding example, OnClickEvent is called directly, without the need for a custom method. MouseEventArgs is passed to the parent component because it is supported by the prebuilt event.

There is another way to use events in the component itself. There can be a situation where you need to update the component's data when an event occurs in the component. You can use Lambda expressions to achieve this update without the need for a custom method. In the following example, we are creating a counter that updates the information about button clicks in a single component:

Counter.razor

```
<p>Number of clicks: @count</p>
<button @onclick="@(e => count + 1)">
Click me to update counter
</button>
@code {
   int count = 0;
}
```

In the preceding code, the `Counter` component will update the `count` field with every click of the button. The e argument in the Lambda expression represents the event argument for the click event: `MouseEventArgs`.

The last type of event that's used in a component is property **two-way binding**. Interactive elements on the website must be able to update themselves. The input element must update the text when the keyboard is pressed but also notify the parent component (for example, the form) that something has happened. Two-way binding is a way to automatically pair the `Parameter` property with the `EventCallback` property. In the following example, two-way binding has been enabled by using an input component with various properties:

CustomInput.razor

```
...
[Parameter]
public string BindingValue
{
get => _value;
set
{
   if ( _value == value ) return;
   _value = value;
   BindingValueChanged.InvokeAsync(value);
}
}

[Parameter]
public EventCallback<string> BindingValueChanged
   { get; set; }
...
```

The `BindingValue` and `BindingValueChanged` properties in the preceding code enable the two-way binding of the `BindingValue` property since there is an `EventCallback` property with the same name as the property, followed by `Changed`. The following component will create an input with two-way binding:

CustomForm.razor

```
...
<CustomInput @bind-BindingValue="@message" />
```

```
...
private string message = "";
```

In the preceding component, when the `BindingValue` property changes in the `CustomInput` component, the parent `CustomForm` component updates the value of the `message` field automatically.

Now that you know how components can communicate, it's time to look at the component life cycle.

Understanding the component life cycle

The life cycle methods for Blazor components are defined in the `BaseComponent` class. Thus, all the components have to inherit – at any level – from the `BaseComponent` class. The `BaseComponent` class has both synchronous and asynchronous life cycle methods. If you need to modify component initialization or rendering, these methods can be overridden. Some methods are only called the first time the component is rendered; others are called every time the component has changed.

The following diagram shows the component life cycle events in order:

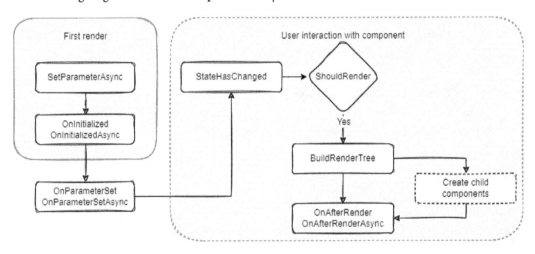

Figure 2.1 – Blazor component life cycle events

In the preceding diagram, you can see the events that are called when the component is initialized from the parent component. The `SetParameterAsync` and `OnInitialized/OnInitializedAsync` methods are only called when the component is rendered for the first time. The re-rendering of the component does not trigger these methods.

After the `OnInitialized` method, the `OnParameterSet/OnParameterSetAsync` methods are called. These methods are called whenever the component parameters change. The change of the method can trigger the re-rendering of the component.

When the `ShouldRender` method returns `true` or is in the first render cycle, the render tree is created. The render tree is then sent to the DOM. In the first render cycle, the render tree needs to be created for the first time, so the `ShouldRender` method is not called in this cycle and the process continues the same way as when the method returns `true`.

After the DOM is updated, the `OnAfterRender`/`OnAfterRenderAsync` methods are called. These methods should not trigger the component UI update since this would lead to the component looping between the `StateHasChanged` and `OnAfterRender` methods.

Structuring component code

Component code in the Razor component can be divided into three sections – the directive section at the top of the file, the section that contains Razor markup, and the C# code with defined properties, fields, and methods at the bottom of the file.

It is common to not need all of the sections in every component file. Our `PageHeader.razor` component only has the middle Razor markup section because there was no need for directives or additional C# code.

Having a lot of methods and functions in the component file can lower the readability of the component itself and can lead to potential mistakes. To prevent this, the `@code` block of the component can be moved to a separate C# file.

As we learned in the previous sections, the Razor component is translated into a C# partial class. To separate the component `@code` from the leading directives and Razor markup, we need to create the C# file with a partial class named the same as the component file. For example, in terms of our `CustomInput.razor` component file, the C# file will be named `CustomInput.razor.cs`.

We can create the file manually, or we can select the `@code` directive in our component. Open the **Quick Action bar** (*Ctrl +* . or *Alt + Enter*) and select **Extract block to code behind**. The Quick Action bar is shown in the following screenshot:

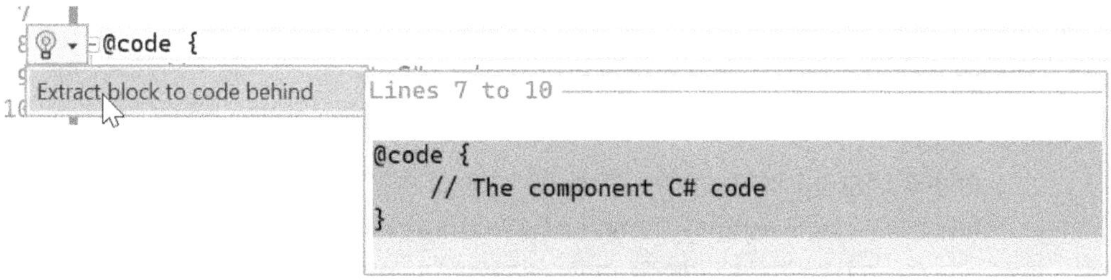

Figure 2.2 – The Quick Action bar

Clicking on **Extract block to code behind**, as shown in the preceding screenshot, will generate a new `.cs` file named `ComponentName.razor.cs`.

The created `public partial class CustomInput` has an inheritance from the `ComponentBase` class automatically, despite the inheritance not being specified in the `.cs` file.

> **Important note**
>
> To specify our custom base class for the component, our base class must inherit from the `ComponentBase` class in the `Microsoft.AspNetCore.Components` namespace. Also, we need to inherit from our base class in the `.cs` file and the `.razor` file using the `@inherits` directive.

The Razor component can be a small part of the website or a whole page with many different components on it. The only difference is in the directive used at the top of the file. Let's look at how we can specify the component's URL address.

Understanding page routing in Blazor

Blazor WebAssembly is a SPA. This means that routing is not done on the server but the client. While clicking on the link updates the address bar of the browser, the page itself is not refreshed. Blazor finds a component with a matching route in the `@page` directive and renders this component (and all child components) as a page.

The `Router` component takes care of resolving the correct component to render. This component is typically used in the application root component – that is, `App.razor`. The component is created from two RenderFragments, as follows:

```
<Router AppAssembly="@typeof(App).Assembly">
    <Found Context="routeData">
        <RouteView RouteData="@routeData"
          DefaultLayout="@typeof(MainLayout)" />
        <FocusOnNavigate RouteData="@routeData"
          Selector="h1" />
    </Found>
    <NotFound>
        <PageTitle>Not found</PageTitle>
        <LayoutView Layout="@typeof(MainLayout)">
            <p role="alert">Sorry, there's nothing at this
              address.</p>
        </LayoutView>
    </NotFound>
</Router>
```

In the preceding code, you can see the Router component with the Found and NotFound sections. The Router component looks to the specified AppAssembly and discovers all routable components (components with the @page directive). When navigating from one page to another, the Router component renders the Found section, if there is any routable component with a matched route. Otherwise, the NotFound section is rendered.

Rendering of the component is handled in the RouteView component. The RouteData parameter of the component contains information about the matched component. RouteView uses a layout defined on the component's @layout directive, or DefaultLayout if the component layout is not specified.

The FocusOnNavigate component is not required. This component sets focus on the element specified by the CSS selector in the Selector parameter. It is useful when building websites compatible with screen readers, or when you want to focus on some user input fields after changing the page.

Navigating between pages

To navigate between pages, the standard HTML anchor is used:

```
<a href="/contacts">Contacts</a>
```

As shown in the preceding code, there is no difference between page navigation in standard HTML and Blazor. That is one of Blazor's advantages over other JavaScript SPA frameworks, which use the custom component to provide SPA navigation. Blazor intercepts any navigation on the same site and tries to find the corresponding components to render.

If you need to manipulate the address from the C# code, you must inject NavigationManager into your code, as follows:

```
@page "/offers"
@inject NavigationManager NavManager
...
<button @onclick="GoToContacts">Contacts</button>
@code {
  void GoToContacts()
  {
    NavManager.NavigateTo("contacts");
  }
}
```

The preceding code will change the page to /contacts after the button is clicked.

Page directive

The @page directive can be specified multiple times in the component. It can also contain dynamic parts of the URL:

PageDirective.razor

```
@page "/author"
@page "/author/{Name}"
<h1>Author @authorName </h1>
@code {
  private string authorName;
[Parameter]
  public string Name { get; set; }

  protected override void OnInitialized()
  {
    authorName = Name ?? "Not set";
  }
}
```

In the preceding code, the first @page directive specifies the URL without any parameters. The second @page directive specifies a parameter with the name. The following URLs are valid for the PageDirective component:

- /author
- /author/John
- /author/123

In the third example, /author/123, you can see that we can pass numbers to the text parameter. It is fine here because any value can be treated like a string. But what if we expect a different type?

Route constraints

Route constraints are used when you need to enforce a specific data type of the route parameter. The route constraints are defined the same way as in C# API endpoints – that is, by adding a semicolon after the parameter name and then specifying the data type:

```
@page "/author/{Id:int}"
```

The preceding code will define the URL for the author with `Id` as an integer.

Not all constraints are supported at the time of writing. The following table shows the supported types:

Constraint type	Example of constraint	Example of valid values
`bool`	`@page "/user/{enabled:bool}`	`True`, `true`, `FALSE`
`datetime`	`@page "/user/{createdAt:datetime}`	`2022-04-23`, `2022-01-01 6:18am`
`decimal`	`@page "/basket/{price:decimal}`	`59.99`, `-999.99`
`double`	`@page "/package/{weight:double}`	`1.15`, `-39.12`
`float`	`@page "/package/{weight:float}`	`1.234`, `-1.234`
`guid`	`@page "/user/{id:guid}`	`905E47D7-DA48-4301-8137-B25541438240`, `{FEB6A90F-3D25-46D1-AD62-64BBD499A0D6}`
`int`	`@page "/user/{id:int}`	`123456`, `-123456`
`long`	`@page "/timer/{ticks:long}`	`123456789`, `-123456789`

Table 2.2 – Blazor router supported constraints

Component route parameters can also be marked as optional:

```
@page "/user/{id:int}/{valid:bool?}"
```

The preceding code specifies the URL to `/user` with the required `Id` parameter and optional `Valid` bool parameter. The optional ? symbol can also be used without type constraints.

Catch-all parameters

Sometimes, you may need to create components for multiple routes. In that case, a catch-all parameter can be used:

Posts/AllPosts.razor

```
@page "/posts/{*pageRoute}"
@code {
```

```
    [Parameter]
    public string? PageRoute { get; set; }
}
```

The preceding code shows a catch-all parameter named `PageRoute`. The catch-all route parameter must be of the `string` type and placed at the end of the route; it is not case-sensitive.

The `PageRoute` parameter will contain all the matched values from the URL:

- For the `/posts/animals/africa/forest` URL, the `PageRoute` value is `animals/africa/forest`

- For the `/posts/animals` URL, the `PageRoute` value is `animals`

- For the `/posts/animals/europa%2Fwater` URL, the `PageRoute` value is `animals/europa/water`

With that, we know how to create components and how to navigate between them. In the next section, we will look at the project we will be creating in this book. And yes! We will use everything we've learned in this chapter up until this point.

Project overview and preparation

In this book, we will be building a complex Blazor WebAssembly application for managing a movie collection. Our application will be able to manage movies and their directors, actors, categories, and more.

We do not want to create an application with lots of functionality at the beginning. Instead, we want to create an application that can be easily extended with other functionality, such as trailers, reviews, and so on.

In *Chapter 3*, *Creating a Database Using Entity Framework Core*, and *Chapter 4*, *Connecting Client and Server with REST API*, we will introduce generic components to communicate between the Blazor application client and server parts. In *Chapter 5*, *Building gRPC Services*, we will learn how to replace the integrated REST API with gRPC communication to improve application performance. In the last part of our project, in *Chapter 6*, *Diving Deep into Source Generators*, we will be implementing source generators to generate repetitive parts of the code. Let's begin with a simple application.

Creating a demo Blazor WebAssembly project

Our demo project is based on the default **Blazor WebAssembly App** template in Visual Studio 2022. This template can be a little bit different, depending on the exact minor version of Visual Studio. After we have used the template to create the project, we will examine its different parts and run the default project to see if everything works as expected. Then, we will modify the existing components and create our components to present list views and details of the *MediaLibrary* data.

Creating the demo MediaLibrary project

Visual Studio has many predefined templates for different kinds of projects. We are going to use the **Blazor WebAssembly App** template for this. This template can be configured differently, so we need to set everything properly. Follow these steps:

1. Open Visual Studio 2022.
2. Click the **Create a new project** button in the right bottom part of the window.
3. Use the **All languages** filter to select **C#**. Then, in the **All project types** filter, select **Blazor** to find **Blazor WebAssembly App**.

    The following screenshot shows the **Blazor WebAssembly App** project template in the Visual Studio 2022 template list:

 Blazor WebAssembly App

A project template for creating a Blazor app that runs on WebAssembly and is optionally hosted by an ASP.NET Core app. This template can be used for web apps with rich dynamic user interfaces (UIs).

C#   Linux   macOS   Windows   Blazor   Cloud   Web

Figure 2.3 – Blazor WebAssembly App project template

4. Select the **Blazor WebAssembly App** template and click the **Next** button.
5. On the **Configure your new project** screen, set the **Project** name to MediaLibrary, specify a **Location** for the project, and click **Next**. The **Solution name** area will be automatically filled with the same project name.

The following screenshot shows the **Configure your new project** window:

Configure your new project

Blazor WebAssembly App   C#   Linux   macOS   Windows   Blazor   Cloud   Web

Project name

MediaLibrary

Location

D:\BlazorProjects   ▼   ...

Solution name ⓘ

MediaLibrary

☐ Place solution and project in the same directory

Back   Next

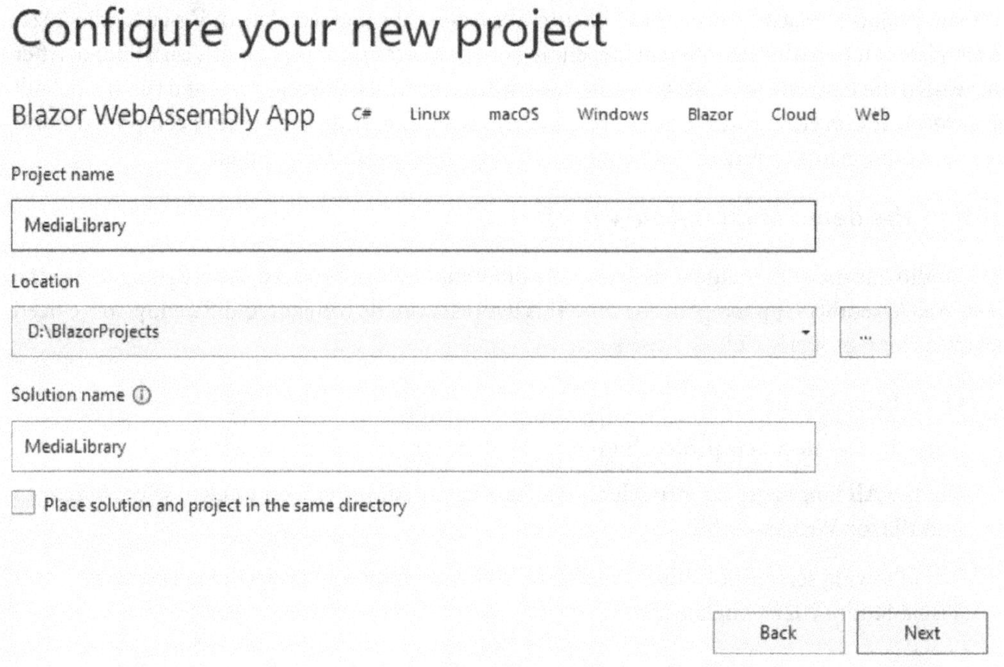

Figure 2.4 – The Configure your new project window

In the preceding screenshot, you can see that we set `D:\BlazorProjects` as our project's **Location**. The location is not important, so you can use whatever location suits you. Click **Next**.

6. In the **Additional information** window, ensure that **Framework** is set to **.NET 6.0 (Long-term support)** and that **Authentication type** is set to **None**. Then, check the checkboxes for **Configure for HTTPS** and **ASP.NET Core hosted**.

The required state is shown in the following screenshot:

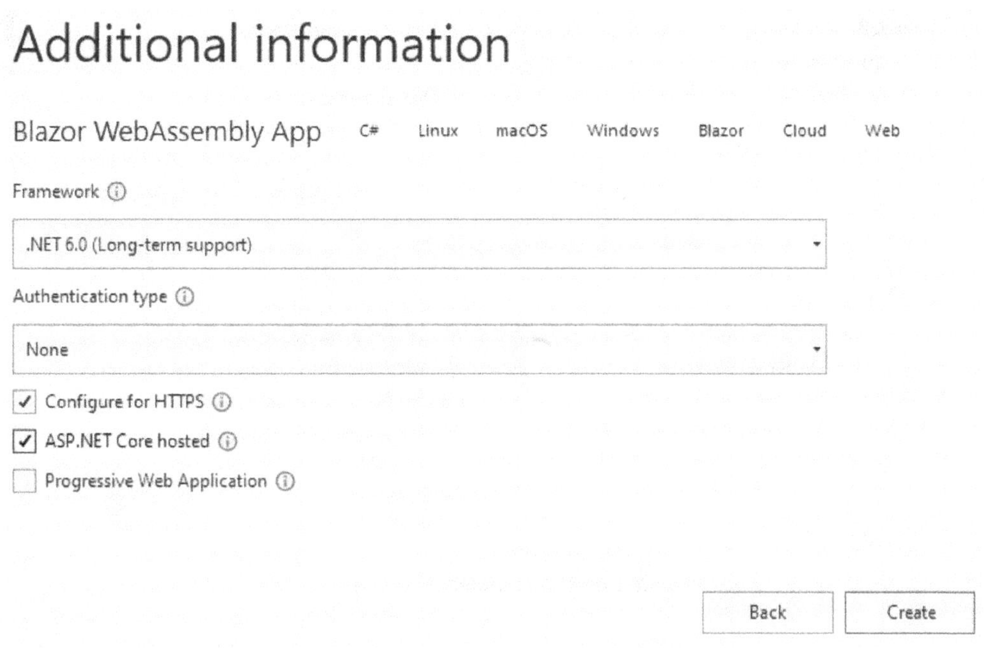

Additional information

Blazor WebAssembly App   C#   Linux   macOS   Windows   Blazor   Cloud   Web

Framework ⓘ

.NET 6.0 (Long-term support)   ▾

Authentication type ⓘ

None   ▾

☑ Configure for HTTPS ⓘ
☑ ASP.NET Core hosted ⓘ
☐ Progressive Web Application ⓘ

Back   Create

Figure 2.5 – Additional information for setting up the Blazor WebAssembly App template

7. Click the **Create** button. The project will be created and opened.

Running the project

Now, you can run the project to see what the project does. The Demo project from the template contains three pages, where each one shows a different functionality. On the **Home** page, there are static texts. The **Counter** page has a **Click me** button and shows how many times the button was clicked. On the **Fetch data** page, there is a component showing the weather forecast that was downloaded from the REST API call to the server.

You can run the project using the **Debug** menu and choosing **Start Without Debugging** or by using the *Ctrl + F5* shortcut. Visual Studio may ask you to install an SSL Certificate to run on HTTPS. Depending on the browser, there may be a warning message about accessing an untrusted website. It is safe to access the site so long as you are on the local host domain.

Examining the project

The advantage of using the Visual Studio **Blazor WebAssembly App** template is that we don't start with an empty project and many requirements to run this type of project are preconfigured.

The following figure shows the generated project's structure:

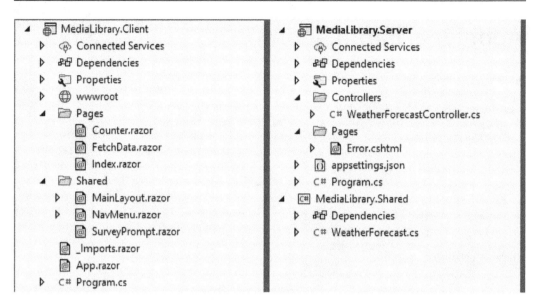

Figure 2.6 – Demo project structure

In the preceding figure, you can see that three projects have been generated in our **MediaLibrary** solution.

Client project

The client project contains our WebAssembly application. It is constructed from a single `.cs` file and multiple Razor components and static files, including images, CSS, JavaScript, and so on.

The wwwroot folder

The `wwwroot` folder is a container for the client application. In our project, you can see that this folder holds custom CSS files, icons, favicon, and an `index.html` file. This folder can be used for public static resources, such as images and other files, which are not built but are published with the application.

The `index.html` file is the mounting point of the application. When the browser requests your application, the `index.html` file is downloaded. This file contains references to additional CSS files, the title of the website, and other meta information. Additional content can be injected into the `<head>` section of the file from the Razor components.

The `<body>` section of the file contains two `div` elements and one `script`:

```
<body>
    <div id="app">Loading...</div>

    <div id="blazor-error-ui">
        An unhandled error has occurred.
        <a href="" class="reload">Reload</a>
        <a class="dismiss">✖</a>
    </div>
    <script src="_framework/blazor.webassembly.js">
        </script>
</body>
```

The `script` element has a link to a file provided by the Blazor framework. This JavaScript file contains all the logic needed to run the WebAssembly application in the browser. The file will download the .NET runtime and your application's assemblies with all their dependencies.

Another highlighted part of the code is the `div` element, which has an `id` attribute with a value of app. This is where all the components will be rendered.

A `blazor-error-ui` element is where unhandled exceptions are shown in case there are any problems on the website.

The Pages folder

The `Pages` folder contains all Razor components that have the @page directive. These components are discovered as routable components.

> **Note**
>
> The routable components can be discovered even outside the `Pages` folder but comprise a common way to separate components in the project. The correct component organization will help you navigate through the files in the project.

The Shared folder

The `Shared` folder contains all the Razor components, which are shared between pages. This folder is a place for all the components that are not routable, regardless of whether you use the component on one page or five. The `Shared` folder can contain additional folder structure to help organize the components.

This folder is also used to define the layout component. The `MainLayout.razor` component is defined in our Demo project. The layout component inherits from the `LayoutComponentBase` class, rather than the `ComponentBase` class:

```razor
@inherits LayoutComponentBase

<div class="page">
    <div class="sidebar">
        <NavMenu />
    </div>

    <main>
        <div class="top-row px-4">
            <a href=https://docs.microsoft.com/aspnet/
            target="_blank">About</a>
        </div>

        <article class="content px-4">
            @Body
        </article>
    </main>
</div>
```

The layout components contain the @Body directive, which is not used in other components. In the runtime, the @Body directive is replaced with the generated page content.

The _Imports.razor file

This file contains the @using directives for namespaces. The namespaces included in this file are automatically included in all Razor components files. The default content of this file has namespaces for HttpClient, basic Razor components provided by the ASP.NET Core team, the JSInterop library, and the namespace of our Shared folder.

The App.razor file

The App.razor component is the root component of the Blazor WebAssembly application. This component uses the Router component to determine what routable page should be rendered. The default layout component is also defined here. The content of the App.razor component was shown in the *Understanding page routing in Blazor* section.

The Program.cs file

This file is the entry point for the Blazor WebAssembly application and contains the configuration for the application:

```
using MediaLibrary.Client;
using Microsoft.AspNetCore.Components.Web;
using Microsoft.AspNetCore.Components.WebAssembly.Hosting;

var builder = WebAssemblyHostBuilder.CreateDefault(args);
builder.RootComponents.Add<App>("#app");
builder.RootComponents.Add<HeadOutlet>("head::after");

builder.Services.AddScoped(sp => new HttpClient {
  BaseAddress = new
   Uri(builder.HostEnvironment.BaseAddress) });

await builder.Build().RunAsync();
```

Here, you can see the code for creating a new builder for our application. After that, two components are attached. First, the App component is attached to the element with the app ID. After that, the HeadOutlet component is attached inside <head> as a last inner element.

The HeadOutlet component is provided by the Blazor framework and is used to change the title of the page or add additional meta information from the rendered components.

Server project

The server project contains part of the application that is executed on the server. It is a generated API project with support for Razor Pages and has a fallback to the index.html file, which is the mounting point for our client WebAssembly application.

This project is no different from other types of WebApi projects. We will use this part to create our REST API, gRPC services, and more.

Shared project

The shared project is referenced by both the client and server projects in our *MediaLibrary* solution. This project is used for the code that needs to be shared between these two projects. It is mostly used to specify **Data Transfer Objects (DTOs)**, models, custom types, and enums.

Preparing the demo project

Now, let's prepare the demo project for our needs. The first thing we have to do is delete all the unnecessary files. So, we need to delete the following files:

- `MediaLibrary.Client`:

  - `Pages/Counter.razor`

  - `Pages/FetchData.razor`

  - `Shared/SurveyPrompt.razor`

- `MediaLibrary.Server`:

  - `Controllers/WeatherForecastController.cs`

- `MediaLibrary.Shared`:

  - `WeatherForecast.cs`

Once we've deleted the files, we should verify that the project is still correct. We can do this by building the project using the **Build** menu option and then choosing **Build Solution** (*Ctrl* + *Shift* + *B*). The project should be built without any errors. The navigation to the `Counter` and `FetchData` components defined in `Shared/NavMenu.razor` component will not work because we've deleted the respective pages. We will fix this later when we add new pages. Now, we have a clean Blazor WebAssembly project for the next chapter.

In this section, we introduced the default **Blazor WebAssembly App** template and explained how to create a demo project using Visual Studio 2022 and this template. Then, we created the demo project and explained the purpose of all the important files. Now, let's summarize this chapter.

Summary

After reading this chapter, you should know about the Razor syntax and how the C# code can be combined with HTML to create dynamic content.

We also covered how to create Razor components, from simple ones containing plain HTML to more advanced components with content defined in parameters by the parent component, as well as the components that can render multiple RenderFragments.

By now, you should know how to use routing in Blazor to specify routable components, navigate between components when the user clicks on the anchor element, and use code behind the `NavigationManager` class. With all this knowledge, you should be able to create any type of Blazor WebAssembly application for websites with dynamic content and JavaScript-like events, but without using any JavaScript libraries or scripts. If you know about the older Microsoft **WebForms** technology, you should be able to migrate such applications to the latest framework using Blazor components.

In the next chapter, we will continue with our demo project. We will create custom routable components that will show a different type of data, create some resource classes that will provide the data for us, and learn how to connect the client with the server to pass the data to the website.

3

Creating a Database Using Entity Framework Core

A Blazor WebAssembly application can be made in the form of a static website without any dynamic data. In such a scenario, only the client part of the application is needed. In our case, we would like to present some data in our application, and therefore we need the server part of the Blazor WebAssembly application. This part will act like an **application programming interface** (**API**) providing data to our client.

In this chapter, we will learn how to create a data repository to store and manipulate data. We will learn about **Entity Framework** (**EF**), which is a great tool for the code-first database approach. We will cover some drawbacks of using EF over **stored procedures**, but also the advantages. We will also explore how to use the EF to create a database in a **Microsoft SQL Server** (**MSSQL**).

We will then learn about *C# generic* to create generic classes and methods that can be reused and can save us a lot of code writing. Later, we will look at **Create, Read, Update, and Delete** (**CRUD**) operations.

By the end of this chapter, you will understand how to connect the *C#* application to the MSSQL server using EF. You will also understand the principles of the generic approach in *C#* and how to use it.

In this chapter, we will cover the following topics:

- Creating a data repository using EF
- Using generic services for data manipulation
- Creating data services
- Registering data services

Technical requirements

All the code for this chapter can be found at https://github.com/PacktPublishing/gRPC-Powered-Blazor-WebAssembly-Development/tree/main/ch3.

Creating a data repository using EF

Each data-driven application needs a place to store data. Using the filesystem can be enough for one application but is not a good option when multiple requests access the same data. This choice is also not the fastest.

Modern applications use databases to store data. These can be relational databases such as *MSSQL*, *MariaDB*, *SQLite*, or *Oracle Database*, or *NoSQL* databases such as *Redis*, *MongoDB*, or *Cassandra*. In our demo project, we will use the *MSSQL database* because it is one of the most commonly used databases for applications written in *C#*.

However, the database itself is the only place where data is stored and comprises a set of mechanisms for how data is handled when we want to create, read, update, or delete any record. Each application needs a connector that will open the communication between the application and the database and transform data from the database to make it readable in the application.

Installing NuGet packages

To extend the functionality of our project, we need to install different NuGet packages. We can install them using *Visual Studio GUI*, *Package Manager Console*, or *Windows Terminal*. In the following examples, we will show the installation of the `Microsoft.EntityFrameworkCore` package to the `MediaLibrary.Server` project using all three options.

Visual Studio GUI

To install the `Microsoft.EntityFrameworkCore` package using Visual Studio *NuGet Package Manager*, proceed as follows:

1. Open the `MediaLibrary` project in Visual Studio.

2. Click the right mouse button on **Dependencies** in the **MediaLibrary.Server** project, as shown in the following screenshot:

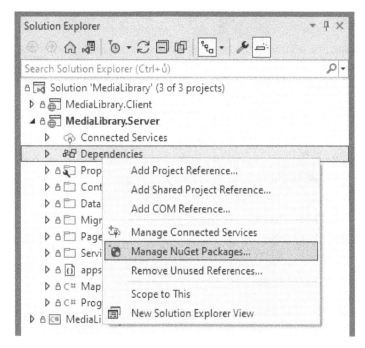

Figure 3.1 – Using NuGet Package Manager

3. Select **Manage NuGet Packages…**. A **NuGet** window will open.

4. Select **Browse** and type EntityFrameworkCore in the search box.

5. Select the **Microsoft.EntityFrameworkCore** package from the list and click the **Install** button.

Package Manager Console

To install the Microsoft.EntityFrameworkCore package using *Package Manager Console*, proceed as follows:

1. Open the MediaLibrary project in Visual Studio.

2. Open the **Package Manager Console** window at the bottom of the screen and select **MediaLibrary. Server** as the **Default project** at top of this window.

3. Run the following command:

```
Install-Package Microsoft.EntityFrameworkCore
```

Windows Terminal

To install the `Microsoft.EntityFrameworkCore` package using *Windows Terminal*, open the terminal (you can also use *PowerShell* or *CMD*) in the folder with the server part of the application (or navigate to the folder) and run the following command:

```
dotnet add package Microsoft.EntityFrameworkCore
```

The result of each of the three preceding approaches is equivalent. The information about package references is added to the project. You can confirm the functionality by opening the `MediaLibrary.Server.csproj` file (right-click on the `MediaLibrary.Server` project and select **Edit Project File**). The file should contain the following line of code:

MediaLibrary.Server.csproj

```
<PackageReference Include="Microsoft.EntityFrameworkCore"
Version="6.0.5" />
```

In the preceding line of code, `PackageReference` is included in the `.csproj` file to notify the project to download and use the specified package in the specific version.

Each approach leads to the same result. It is up to you—choose whichever method suits you best. For the purpose of this book, we will be using the terminal approach, as it is an easier way to install NuGet packages, knowing the exact name of the package.

Enabling EF

EF Core is a popular open source, cross-platform data access technology. EF Core is used as an **object-relational mapper (ORM)** that enables us to work with a database using *.NET* objects and eliminates the need for writing a stored procedure.

> **What is a stored procedure?**
>
> A stored procedure (sometimes called SP) is a set of SQL code with an attached name, saved in the database. This code can be then called from other SQL code or C# applications. Stored procedures comprise reusable code that can be parametrized to return different results or to store different data.

EF Core allows us to generate an object model of our database, or—in our case—write the object model in C# and use **migrations** to create a database or update the database model.

The main advantage when using EF Core is the speed of development. Using standard stored procedures, the database model must be designed with different tools and languages. Also, all operations on the database have to have their own stored procedures, which leads to a large number of stored procedures in the project.

On the other hand, there are also some disadvantages. EF Core will generate SQL queries from our C# **Language-Integrated Query (LINQ)** code. If the LINQ code is simple, the generated SQL will perform well enough. However, when you need to retrieve complex data with multiple joins, the generated SQL code can perform worse than in a manually written stored procedure.

> **Note**
>
> Many applications use multiple ORMs at the same time. EF is one of the best ORMs for a system with administration, where standard CRUD operations are required. When requesting complex data from the database, different ORMs can be in place. One of the most common is *Dapper*. Dapper can be used to map a stored procedure to the C# class and provides a smooth C# experience in terms of the performance of stored procedures.

To use EF Core in our project, we need to install the required NuGet packages first.

Here are the required packages:

- `Microsoft.EntityFrameworkCore`

  We have already installed this package to the project.

- `Microsoft.EntityFrameworkCore.SqlServer`

  To install this package, open the terminal in `path_to_project\MediaLibrary\Server` and run the following command:

  ```
  dotnet add package Microsoft.EntityFrameworkCore.
  SqlServer
  ```

  The preceding package contains a provider to communicate between the C# application and MSSQL Server.

After the installation, we can confirm that the packages are installed by inspecting the `MediaLibrary.Server.csproj` file, as follows:

MediaLibrary.Server.csproj

```
<Project Sdk="Microsoft.NET.Sdk.Web">
  <PropertyGroup>
    <TargetFramework>net6.0</TargetFramework>
    <Nullable>enable</Nullable>
```

```xml
      <ImplicitUsings>enable</ImplicitUsings>
  </PropertyGroup>

  <ItemGroup>
    <PackageReference Include="Microsoft.AspNetCore.Components.
    WebAssembly.Server" Version="6.0.5" />
    <PackageReference Include="Microsoft.EntityFrameworkCore"
    Version="6.0.5" />
    <PackageReference Include="Microsoft.EntityFrameworkCore.
    SqlServer" Version="6.0.5" />
  </ItemGroup>

  <ItemGroup>
    <ProjectReference Include="..\Client\MediaLibrary.Client.
    csproj" />
    <ProjectReference Include="..\Shared\MediaLibrary.Shared.
    csproj" />
  </ItemGroup>

  <ItemGroup>
    <Folder Include="Controllers\" />
  </ItemGroup>
</Project>
```

The preceding code block shows the whole .csproj file of our server project. In the beginning, the file specifies TargetFramework as net6.0. Then, we have two flags to enable Nullable and ImplicitUsings. These two options help us write better code.

Then, we have ItemGroup with PackageReference lines for each NuGet package added to our project. In the next group, there is a ProjectReference line for each referenced project. The last group contains information about the included folder.

We now have everything set up to create our first data repository.

Creating database representation in C#

DbContext is a C# class provided by the Microsoft.EntityFrameworkCore package that contains all information about our connection to the database and database model. This is our entry point to create and configure database logic.

Let's create a DbContext class file called MediaLibraryDbContext.cs. We will create it in the MediaLibrary.Server project and add the inheritance of the DbContext class, as follows:

Data\MediaLibraryDbContext.cs

```
using Microsoft.EntityFrameworkCore;
namespace MediaLibrary.Server.Data;
public class MediaLibraryDbContext : DbContext
{


}
```

The preceding code snippet shows the full content of the created file. Our MediaLibraryDbContext class must inherit from Microsoft.EntityFrameworkCore.DbContext. Since we have the using directive of the required package, only the DbContext name is used in the inheritance specification.

Now, we can create our first database entity. Let's start with a simple Person entity. We will create new files in MediaLibrary.Server, like so:

Data\BaseEntity.cs

```
using System.ComponentModel.DataAnnotations;
namespace MediaLibrary.Server.Data;
public class BaseEntity
{
    [Key]
    public int Id { get; set; }
    public string Name { get; set; } = string.Empty;
}
```

The preceding code will create a class called BaseEntity. This class will contain Id and Name properties, and we will implement this class in almost all of the entities we are going to create. The Key attribute on the Id property notifies EF Core that this property is the **primary key (PK)** in the database. The code is illustrated in the following snippet:

Data\Person.cs

```
namespace MediaLibrary.Server.Data;
public class Person : BaseEntity
```

```
{
    public DateTime BirthDay { get; set; }
    public string BirthPlace { get; set; } = string.Empty;
    public string Biography { get; set; } = string.Empty;
}
```

The preceding code creates a `Person` class inheriting the `BaseEntity` class. We now have a `Person` entity that will have `Id`, `Name`, `BirthDay`, `BirthPlace`, and `Biography` columns in a database.

Right now, EF Core does not know about our `Person` class. We need to notify EF Core to use this class and map it to a generated database table. There are two ways of doing that: we can mark the class with the `Table` attribute in the same way as we marked the `Key` attribute, or we can use **Fluent API** to configure the entities in the `MediaLibraryDbContext` class. In this book, we will be using Fluent API for the EF Core configuration (except for the `Key` attribute).

In the `MediaLibraryDbContext` class, we need to override the `OnModelCreating` method and create a `public DbSet` property for the `Person` entities, as follows:

Data\MediaLibraryDbContext.cs

```
...
public DbSet<Person> Persons { get; set; }

protected override void OnModelCreating(ModelBuilder
    modelBuilder)
{
    modelBuilder.Entity<Person>(b =>
    {
        b.Property(x => x.Name).IsRequired();
    });
}
...
```

In the preceding code snippet, we are creating a public `DbSet<Person>` property: `Persons`. This property will create access to all records stored in the `Person` table. Next, we are overriding the `OnModelCreating` method with our implementation.

Our implementation contains the registration of the `Person` entity and marks the `Name` property as `Required`.

We also need to specify constructors for our `MediaLibraryDbContext` class. Add the following code at the beginning of the class:

```
public MediaLibraryDbContext() { }
public
MediaLibraryDbContext(DbContextOptions<MediaLibraryDbContext>
options) : base(options) { }
```

The preceding code creates two constructors. The empty one is the default, and the second one is used when a database is generated from our code.

Right now, we have configured EF Core to create a database table for the `Person` entity and to allow us to manipulate the records in this table. Now, we need to configure our `MediaLibrary` Blazor application to use the EF Core ORM.

Configuring EF Core

The configuration of EF Core is simple, and you need just a few lines of code. Open the `Program.cs` file and add two `using` directives to the top of the file, as follows:

```
using MediaLibrary.Server.Data;
using Microsoft.EntityFrameworkCore;
```

The preceding code will allow us to use the methods from the `Microsoft.EntityFrameworkCore` package and our `Data` folder.

Next, add the following code just after the `builder` declaration:

```
...
builder.Services.AddDbContext<MediaLibraryDbContext>(options =>
{
    options.UseSqlServer(builder.Configuration.
GetConnectionString("MediaLibrary"));

#if DEBUG
    options.EnableDetailedErrors();
    options.EnableSensitiveDataLogging();
#endif
});
...
```

The preceding code will configure our application to use EF Core with our `MediaLibraryDbContext` class as the database context. It also configures EF Core to use MSSQL as a database, with a connection string named `MediaLibrary` from our configuration file. The code is illustrated in the following snippet:

appsetting.Development.json

```json
{
  "Logging": {
    "LogLevel": {
      "Default": "Information",
      "Microsoft.AspNetCore": "Warning"
    }
  },
  "ConnectionStrings": {
    "MediaLibrary": "Data source=SERVER_NAME;Initial
      catalog=MediaLibraryDb;Integrated Security=True; "
  }
}
```

The preceding code is the full content of the `appsettings.Development.json` file that holds our local configuration. `SERVER_NAME` in the `ConnectionStrings.MediaLibrary` should be changed to your server's name.

Migrating code to the database

When we add, change, or delete any entity, we need to update the database schema. Such updating processes are called migrations, and each changeset is called a migration. There are multiple ways of achieving this outcome, but we will use terminal commands. To be able to use `dotnet ef` commands, we need to install the `Microsoft.EntityFrameworkCore.Design` package first, like so:

```
dotnet add package Microsoft.EntityFrameworkCore.Design
```

To create a migration, we need to run the following command in the terminal:

```
dotnet ef migrations add InitialMigration
```

The preceding command will generate two files in the `Migrations` folder. The `MediaLibraryDbContextModelSnapshot.cs` file is a snapshot of current state. It is a representation of what the database should look like. The other file consists of a timestamp and our migration name. This file contains two methods: the Up method is used when the migration is transferred to the SQL database, and the Down method is used when we want to revert the migration from the server.

When we create other migrations, the `MediaLilbraryDbContextModelSnapshot.cs` file will be updated, and a new migration file will be added.

The last thing we need to do to have a working database for our project is to apply migrations to the SQL server. This can be achieved by running the following command:

```
dotnet ef database update
```

The preceding command connects the application to the database, compares existing migrations, and applies the ones that are not on the SQL server.

Creating entities in the database

Now that we know how to create C# classes as database entities, let's create them the way we will need them to be in our demo project.

Person

The `Person` entity will contain information about one single person. This person can be a director, actor, music composer, and so on. Since most of the persons will be actors, the `Person` entity should contain a collection of movies. We do not have the `MovieActor` class yet, so you can ignore the error provided by Visual Studio IDE. The following code snippet shows the full contents of the file:

Data\Person.cs

```
namespace MediaLibrary.Server.Data;
public class Person : BaseEntity
{
    public DateTime BirthDay { get; set; }
    public string BirthPlace { get; set; } = string.Empty;
    public string Biography { get; set; } = string.Empty;
    public List<MovieActor> Movies { get; set; } = new
        List<MovieActor>();
}
```

The `Movies` property is added to the `Person.cs` file. This collection will be used to attach the movies where the person was an actor, to the person.

Category

The Movie category is a simple list of categories. We can define them as a class, translated to the database entity, or we can use the advantages of C# and define them as an enum. Because we will need to use the enum in both the server and client parts of the application, we need to define it in the MediaLibrary.Shared project, like so:

MediaLibrary.Shared\CategoryType.cs

```
namespace MediaLibrary.Shared;
public enum CategoryType
{
    Action = 0,
    Comedy = 1,
    Drama = 2,
    Fantasy = 3,
    SciFi = 4,
    Horror = 5,
    Mystery = 6,
    Romance = 7,
    Thriller = 8,
    Western = 9
}
```

The preceding code snippet shows the definition of the enum in C#. Each value of the enum is translated to the relevant int value. When the enum value is stored in the database using EF Core, we can configure the application to store it as an int representation, or a string value.

> **Note**
>
> The numeric value of the enum does not need to be specified. The enum values start from 0 and follow the n+1 schema. If we specify the first value, the next will follow the same schema. If we specify the value in the middle of the enum, all the following values will continue with the numbering from that specified value.
>
> When we specify values by hand, we are suggesting to the other developers that they should not change the numeric values of the enum.

Each approach has its benefits and drawback. Storing as `int`, we need persistent `int` values for the `string` representation. The database value is smaller and faster to read/write and filter. Using `string` values is slower, and we can't modify the existing `enum` values (for example, if we have a typo in the value). On the other hand, a `string` value is readable on the database level.

Each movie should ideally belong to one category, such as drama, horror, comedy, and so on. However, in reality, some movies can belong to several categories. But since we can't store an `array` or a `list` in the same database column, we need to create another entity to store multiple categories for a single movie, as follows:

Data\MovieCategory.cs

```
using MediaLibrary.Shared;
namespace MediaLibrary.Server.Data;
public class MovieCategory
{
    public int MovieId { get; set; }
    public Movie Movie { get; set; } = null!;
    public CategoryType Category { get; set; }
}
```

The preceding code snippet shows the full content of the `MovieCategory.cs` file. This class will create a binding table in the SQL server. This is a table used to deconstruct **many-to-many (M:N) relations** between two entities.

You can find more about **one-to-one (1:1)**, **one-to-many (1:N)**, and M:N relations via the links provided at the end of this chapter.

Here, we have `CategoryType` on one side and `Movie` on the other. Each category can have multiple movies, and each movie can be in multiple categories. The *M:N* relation is not possible to create in the **relational database system (RDBS)**, so we had to deconstruct it into two *1:N* relations.

> **Note**
> NoSQL databases store data in different ways (for example, in the JSON format) and allow the storing of array/list entities.

The `MovieCategory` entity also has two properties pointing to the `Movie` entity. The first one is `MovieId` and the second is called `Movie`. `MovieId` represents the column created in the database and holds relation to the `Movie` entity. The `Movie` property then contains the entire `Movie` entity, when loaded from the database.

> **Important information**
>
> When you name an object property with `SomeName`, a `SomeNameId` property should be created. EF Core then knows that these two properties point to a single item in the database.

Movie

Movies are the main purpose of the demo application. We need an entity that will store all data about our movies. The following code snippet shows the full contents of the `Movie.cs` file:

Data\Movie.cs

```
namespace MediaLibrary.Server.Data;
public class Movie : BaseEntity
{
    public List<MovieCategory> Categories { get; set; } = new
        List<MovieCategory>();
    public int Year { get; set; }
    public string? Description { get; set; }
    public Person? Director { get; set; }
    public int? DirectorId { get; set; }
    public Person? MusicComposer { get; set; }
    public int? MusicComposerId { get; set; }
    public List<MovieActor> Actors { get; set; } = new
        List<MovieActor>();
}
```

The preceding code snippet shows the whole definition of the `Movie` entity. Note that this entity also inherits from the `BaseEntity` class to contain the PK and name. Our movie also contains a collection of categories, the year when the movie was released, and a description. Next, the movie has a relation to multiple persons such as the movie director, the music composer, and a list of actors.

The director and music composer are specified with two properties: the `Id` property for creating the navigation between `Person` and `Movie`, and the `Person` type property to hold `Director` or `MusicComposer`.

Because the movie has multiple actors, and actors play in multiple movies, we need to create a binding table for the actors in the movie. Here's how we can do this:

Data\MovieActor.cs

```
namespace MediaLibrary.Server.Data;
public class MovieActor
{
    public int MovieId { get; set; }
    public Movie Movie { get; set; } = null!;
    public int PersonId { get; set; }
    public Person Person { get; set; } = null!;
}
```

The preceding code snippet shows the full `MovieActor.cs` file that contains *1:N* relations to the `Movie` entity and the `Person` entity.

Once we have defined our classes, let's tell the EF Core `MediaLibraryDbContext` class to use them as entities.

First, we want to use the `Movie` entity as a collection available in the `DbContext` class. We have already defined a `DbSet<Person>` property called `Persons`, so we can add the following line of code just after that:

```
public DbSet<Movie> Movies { get; set; }
```

The preceding line of code will add `Movies` as an available data endpoint to the `DbContext` class.

Next, we should extend the `OnModelCreating` method one more time, like so:

```
protected override void OnModelCreating(ModelBuilder
modelBuilder)
{
    modelBuilder.Entity<Person>(b =>
    {
        b.Property(x => x.Name).IsRequired();
        b.Navigation(x => x.Movies).AutoInclude(); // Added
            line
    });

    // Added method content
```

```
modelBuilder.Entity<Movie>(b =>
{
    b.Property(x => x.Name).IsRequired();
    b.HasOne(x => x.Director).WithMany().HasForeignKey(x =>
      x.DirectorId).OnDelete(DeleteBehavior.Restrict);
    b.HasOne(x => x.MusicComposer).WithMany().
      HasForeignKey(x => x.MusicComposerId).
      OnDelete(DeleteBehavior.Restrict);
    b.Navigation(x => x.Director).AutoInclude();
    b.Navigation(x => x.MusicComposer).AutoInclude();
    b.Navigation(x => x.Actors).AutoInclude();
});

modelBuilder.Entity<MovieCategory>(b =>
{
    b.HasKey(x => new { x.Category, x.MovieId });
});

modelBuilder.Entity<MovieActor>(b =>
{
    b.HasKey(x => new { x.MovieId, x.PersonId });
});
}
```

The preceding code block shows how to add entities to the DbContext class so that EF Core will know how to handle data and how to generate migrations to the database.

The first entity added to the context is the Movie entity. As in the previous Person entity, we are specifying the Name value as required. For both Person and Movie entities, we are configuring navigation properties to include automatically when selected. We are also creating a Navigation property, saying that the Movie entity can have one Director entity (and the Director entity can have multiple Movie entities) with DirectorId as a **foreign key (FK)**. We are also saying that deleting the Person entity used as Director is restricted. The MusicComposer entity is configured the same way as Director.

Then, we are configuring the MovieCategory and MovieActor entities. These entities are used as a binding table, thus they don't inherit from the BaseEntity class and they don't have a PK specified. For the binding table, the PK should be computed from all of the FKs to create a unique value for each possible record in the table.

Now, it's time to create another migration of the database schema. Run the following command in the `MediaLibrary\Server` folder:

```
dotnet ef migrations add Entities
```

The preceding command will generate a new file in the `MediaLibrary\Server\Migrations` folder. This filename will end with `_Entities.cs`, and the content of the file will contain the migration script.

To modify our database, run the following command:

```
dotnet ef database update
```

The preceding command will compare the database and the project entities and apply the missing migration.

We can confirm that the migration was applied by checking the database schema, as follows:

1. Open **SQL Server Management Studio (SSMS)**.
2. Connect to the `localhost` server.
3. Navigate to the `MediaLibrary` database.
4. Open **Database Diagrams** and create a new diagram.
5. In the **Add Table** window, select all tables and click **Add**.
6. Click the **Close** button.

The generated schema should look like this:

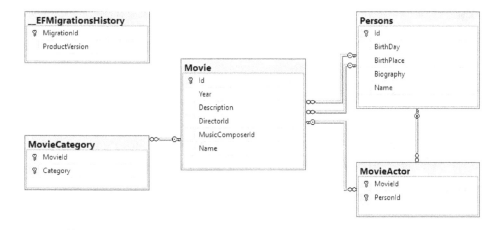

Figure 3.2 – Database schema

The preceding screenshot shows the database schema of the application with all relations between tables.

You now know how to create C# classes as database entities for EF Core, how to create relations between the entities, how to create migrations with changes, and how to apply the migrations to the database. It is now time to create some service classes to be able to CRUD the data.

Using generic services for data manipulation

Each time we need to CRUD some record from the database, we will need to use our `MediaLibraryDbContext` class and write a LINQ query to retrieve the required data. To prevent typing the same lines of code, we will create a service class that will retrieve the data for us.

For reasons including concerns around security, we also should never expose database entities to the client. We can avoid this behavior by creating models that will be shared between the client and server part of the application. The client will only know about the public models, and not about database entities. The service class will CRUD the entities created in the previous section, but the caller of this method will only see the model as method parameters or result values. This can prevent unrequired modifications of the entities and create a more secure and extensible application.

Creating models

In this section, we will create two classes in the `MediaLibrary.Shared` project. We will also create one interface called `IModel` describing required data for all classes, and then models for `Movie` and `Person` entities. The code is illustrated in the following snippet:

Model\IModel.cs

```
namespace MediaLibrary.Shared.Model;
public interface IModel
{
    int Id { get; set; }
    string Name { get; set; }
}
```

The preceding code snippet shows the `IModel` interface that will be our contract for all models where we expect the `Id` and `Name` properties.

Next, we need to create a model class for the `Movie` entity. The `MovieModel` class implementing the `IModel` interface will copy the fields from the entity with just small modifications, as follows:

Model\MovieModel.cs

```
namespace MediaLibrary.Shared.Model;
public class MovieModel : IModel
{
    public int Id { get; set; }
    public string Name { get; set; } = string.Empty;
    public List<CategoryType> Categories { get; set; } =
      new List<CategoryType>();
    public int Year { get; set; }
    public string? Description { get; set; }
    public int? DirectorId { get; set; }
    public int? MusicComposerId { get; set; }
    public int[] ActorIds { get; set; } =
      Array.Empty<int>();
}
```

The preceding code snippet shows the `MovieModel` class for the `Movie` entity. The difference here is that we don't use the `Person` entity and that the collection of `Actor` entities holds the **identifiers (IDs)** of actors (the `Person` entity). In C#, we can have an *M:N* relation between objects, so we skipped the binding table.

The last model is the `PersonModel` class implementing the `IModel` interface, as follows:

Model\PersonModel.cs

```
namespace MediaLibrary.Shared.Model;
public class PersonModel : IModel
{
    public int Id { get; set; }
    public string Name { get; set; } = string.Empty;
    public DateTime BirthDay { get; set; }
    public string BirthPlace { get; set; } = string.Empty;
    public string Biography { get; set; } = string.Empty;
    public int[] MoviesIds { get; set; } =
```

```
        Array.Empty<int>();
}
```

The preceding code shows the `PersonModel` class. The only difference between this class and the `Person` entity is that the model class has a collection of `Movies` IDs as an `int` array.

Once we have defined models for our application, we need to create some mappings between the entity and model. We don't want to assign each entity property from each model property every time we want to manipulate them. To help with this task, the *AutoMapper* NuGet package can be used to configure the mapping only once, and this mapping can be used whenever we want to create a model from an entity and vice versa.

Mappings between models

Before we create mappings, we need to install the *AutoMapper* NuGet packages. Run the following commands in the folder with the server part of the project:

```
dotnet add package AutoMapper
dotnet add package AutoMapper.Extensions.Microsoft.
DependencyInjection
```

The preceding commands will install the required NuGet packages to our server project application.

Next, we need to configure our application to use this package by adding the following line to the `Program.cs` file:

```
builder.Services.AddAutoMapper(typeof(MapperProfile));
```

The preceding code should be added to the `Program.cs` file after the other service registrations (line *20* in the example code).

And finally, we need to create a `MapperProfile.cs` class, as follows:

```csharp
using AutoMapper;
namespace MediaLibrary.Server;
public class MapperProfile : Profile
{
    public MapperProfile()
    {
        // TODO: Person mappings
        // TODO: Movie mappings
    }
}
```

The preceding class will inherit from the `AutoMapper.Profile` class and contain all the configurations for the mappings between our entities and models. The code shows a basic `MapperProfile.cs` file. As you can see, the file has an empty constructor where we will be implementing the mappers. The mappers will be registered during the application startup, so they will not do any dynamic operation. The mappings should be static code that just points from one object to another.

> **Important information**
>
> Properties with the same name and type are mapped automatically. You need to specify only the differences, or when you want to change the default behavior.

Let's create mappings for the `Person` entity and `PersonModel`. The following code snippet shows two mappings: from `Person` to `PersonModel` and vice versa:

```
CreateMap<Data.Person, Shared.Model.PersonModel>()
    .ForMember(x => x.MoviesIds, map => map.MapFrom(actor
      => actor.Movies.Select(x => x.MovieId)));


CreateMap<Shared.Model.PersonModel, Data.Person>()
    .ForMember(x => x.Id, map => map.Ignore())
    .ForMember(x => x.Movies, map => map.MapFrom((model,
      entity) =>
    {
        var current = entity.Movies.Select(m =>
          m.MovieId).ToArray();
        var movies = entity.Movies.ToList();
        movies.AddRange(model.MoviesIds.Where(x =>
          !current.Contains(x)).Select(x => new
          Data.MovieActor
          {
              MovieId = x,
          }));


        movies.RemoveAll(x =>
          !model.MoviesIds.Contains(x.MovieId));


        return movies;
    }));
```

In the preceding code snippet, when we bind an entity to a model, we need only to specify the `MovieIds` collection to be filled from the binding table.

In the mapping from the model to the entity, we want to ignore the `Id` property. The reason for this is that the client should never modify the `Id` property of an entity. Next, we want to merge the actors' collection from the model with the actors in the entity. Adding missing values and removing records that are not in the model will keep existing values intact.

The mappings for the `Movie` entity and `MovieModel` will be almost the same, except for the fact that we need to create mappings for the `Movie` category as well. The following code snippet shows how to do that:

```
CreateMap<Data.Movie, Shared.Model.MovieModel>()
    .ForMember(x => x.Categories, map => map.MapFrom(movie
      => movie.Categories.Select(x => x.Category)))
    .ForMember(x => x.ActorIds, map => map.MapFrom(movie =>
      movie.Actors.Select(x => x.PersonId)));


CreateMap<Shared.Model.MovieModel, Data.Movie>()
    .ForMember(x => x.Id, map => map.Ignore())
    .ForMember(x => x.Categories, map =>
      map.MapFrom((model, entity) =>
    {
        var current = entity.Categories.Select(x =>
          x.Category).ToArray();
        var categories = entity.Categories.ToList();
        categories.AddRange(model.Categories.Where(x =>
          !current.Contains(x)).Select(x => new
          Data.MovieCategory
          {
              Category = x
          }));
        categories.RemoveAll(x =>
          !model.Categories.Contains(x.Category));
        return categories;
    }))
    .ForMember(x => x.Actors, map => map.MapFrom((model,
      entity) =>
    {
```

```
    var current = entity.Actors.Select(x =>
      x.PersonId).ToArray();
    var actors = entity.Actors.ToList();
    actors.AddRange(model.ActorIds.Where(x =>
      !current.Contains(x)).Select(x => new
      Data.MovieActor
      {
        PersonId = x
      }));

    actors.RemoveAll(x =>
      !model.ActorIds.Contains(x.PersonId));

    return actors;
  }));
```

The preceding code snippet shows the mapping for `Movie` and `MovieModel` with specified mappings for `Category` and `ActorIds`, ignoring the `Id` property when mapping from the model to the entity.

> **Note**
>
> If mappings are similar in both ways, you can configure just one way and use the `ReverseMap` extension method to configure mapping the opposite way.

Once we can map between an entity and the corresponding model, we can use these mappings to retrieve an entity from EF Core and return the model to the caller.

Creating data services

In the real world, we will probably continue with creating an `interface` as a descriptor of the methods available in our services. However, for the purpose of this book, the approach without interfaces will be better, as it will lower the amount of code needed to run the application, without any drawbacks. If we need to use interfaces in the future, we can generate them using Visual Studio refactoring tools at any time.

The first service class that we will create will be a generic `BaseService` class that will handle all our database communication:

Services/BaseService.cs

```csharp
using AutoMapper;
using MediaLibrary.Server.Data;
using MediaLibrary.Shared.Model;
using Microsoft.EntityFrameworkCore;
namespace MediaLibrary.Server.Services;
public abstract class BaseService<TEntity, TModel>
    where TEntity : BaseEntity
    where TModel : IModel, new()


{

    private readonly MediaLibraryDbContext _dbContext;
    private readonly IMapper _mapper;

    public BaseService(MediaLibraryDbContext dbContext,
      IMapper mapper)
    {
        _dbContext = dbContext;
        _mapper = mapper;
    }


    // Public methods go here

}
```

The preceding code snippet shows the structure of our base abstract service implementation. We are defining a generic class with two generic parameters. `TEntity` represents our entity model and must be the `BaseEntity` class (or a class that inherits from `BaseEntity`). `TModel` represents our model class and must implement the `IModel` interface. The `new()` part in the `where` clause specifies that `TModel` must have an empty constructor.

The constructor of the `BaseService` class also specifies that the parent class has to inject the `MediaLibraryDbContext` class and the `IMapper` interface.

This service does nothing so far, so we should implement some methods.

Creating a new record

To create a new record, we should implement the CreateAsync method, as follows:

```
public async Task<TModel> CreateAsync(TModel model)
{
    var entity = _mapper.Map<TEntity>(model);
    await _dbContext.Set<TEntity>().AddAsync(entity);
    await _dbContext.SaveChangesAsync();
    return _mapper.Map<TModel>(entity);
}
```

The preceding code snippet shows the implementation of the CreateAsync method. The method creates a new entity from the provided model and then adds the entity to the database context. The SaveChangesAsync extension method will save the modifications in the DbContext class to the database. At this moment, the entity has been updated. As result, the updated entity is mapped back to the model and returned to the caller.

Reading a single record

Reading a single record from the database can be done easily. However, reading a record has to be done not only when we want to return a record to the caller, but also when we update or delete a record. And because we don't want to return the entity to the caller, we should create a private method that we can reuse, as follows:

```
private async Task<TEntity> GetEntityByIdAsync(int id)
{
    var entity = await _dbContext.FindAsync<TEntity>(id);

    if (entity == null)
    {
        throw new Exception($"Cannot find entity type
            {typeof(TEntity)} with id {id}");
    }

    return entity;
}
```

The preceding code snippet shows the GetEntityByIdAsync method that can read records from the database. If the record does not exist, the method throws an exception.

We can now use this method in the public method exposed to the client, as follows:

```
public async Task<TModel> GetByIdAsync(int id)
{
    if (id <= 0)
    {
        return new TModel();
    }

    var entity = await GetEntityByIdAsync(id);
    return _mapper.Map<TModel>(entity);
}
```

The preceding code snippet shows the GetByIdAsync method to retrieve the model from the entity with a specified ID. If the ID is 0, the newly created model is returned.

Reading all records

When reading all values from the database table, we do not expect that the records will be modified. We can use the AsNoTracking method to prevent DbContext from tracking changes on returned entities. This can help with preventing unnecessary memory allocation. You can see an illustration of this method in the following code snippet:

```
public async Task<IEnumerable<TModel>> GetAllAsync()
{
    var entities = await _dbContext
        .Set<TEntity>()
        .AsNoTracking()
        .ToListAsync();
    return entities.Select(x => _mapper.Map<TModel>(x));
}
```

The preceding code snippet shows the GetAllAsync method to retrieve all records from the table in the database. Then, we map the entities to the model and return them to the caller.

> **Important information**
>
> The `ToListAsync` method without any `Select` clause will retrieve all the columns from a table. In our demo, we are using all the columns, but in other applications, you may need to return just a small amount of the columns. Returning unnecessary data can slow down your application. *AutoMapper* has a `ProjectTo` extension method that can be used to retrieve only columns needed by the mapping configuration and return a collection of data mapped from the entity to the model.

Updating records

When updating data, we don't want to create a new entity. Instead, we want to update existing ones. We can achieve this outcome by using the following method:

```
public async Task<TModel> UpdateAsync(int id, TModel model)
{
    var entity = await GetEntityByIdAsync(id);
    _mapper.Map<TModel, TEntity>(model, entity);
    await _dbContext.SaveChangesAsync();
    return _mapper.Map<TModel>(entity);
}
```

The preceding code shows the `UpdateAsync` method. In this method, the entity with the provided `id` property is retrieved from the database; then, the mapping from the model to the entity is done on it. The updated entity is then saved to the database.

Deleting records

Last but not least, we may want to delete some records from the database. We can achieve this by using the following code:

```
public async Task DeleteAsync(int id)
{
    var entity = await GetEntityByIdAsync(id);
    _dbContext.Set<TEntity>().Remove(entity);
    await _dbContext.SaveChangesAsync();
}
```

The preceding code snippet shows the `DeleteAsync` method, which retrieves an entity from the database, marks that entity as removed, and deletes it from the database using the `SaveChangesAsync` extension method.

Registering data services

We have created a data service for our database calls. We now need to register it. But we have created the service as an abstract class, which means we can't use the class itself. We need to implement other services that will inherit from our `BaseService<TEntity, TModel>` class.

First, we can create a service for the `Movie` entity, like so:

Services\MovieService.cs

```
using AutoMapper;
using MediaLibrary.Server.Data;


namespace MediaLibrary.Server.Services;


public class MovieService : BaseService<Movie,
  Shared.Model.MovieModel>
{
    public MovieService(MediaLibraryDbContext dbContext,
      IMapper mapper) : base(dbContext, mapper)
    {
    }
}
```

The preceding code snippet shows the whole file for the `MovieService` class. The class has an inheritance for the `BaseService` class, where we specify `Movie` and `MovieModel` as the generic parameters. Then, a constructor is created to pass parameters to the base class.

Now, we can repeat the same process for the `Person` entity, as follows:

Services\PersonService.cs

```
using AutoMapper;
using MediaLibrary.Server.Data;


namespace MediaLibrary.Server.Services;
public class PersonService : BaseService<Person,
  Shared.Model.PersonModel>
{
    public PersonService(MediaLibraryDbContext dbContext,
```

```
        IMapper mapper) : base(dbContext, mapper)
    {
    }
}
```

The preceding code snippet shows the whole file for the `PersonService` class.

Having these two classes, we can extend our application to use data services. Add the following code after line *20* in the `Program.cs` file:

```
builder.Services.AddTransient<MovieService>();
builder.Services.AddTransient<PersonService>();
```

The preceding code snippet shows how to register custom services to be used in our application.

The `using` directive for `MediaLibrary.Server.Services` should be added automatically. If not, add the following line to the top of the `Program.cs` file, where other `using` directives are:

```
using MediaLibrary.Server.Services;
```

The preceding code snippet shows how to specify a namespace in the `Program.cs` file to use our service classes.

To validate that the created code is correct, we can build our application. It should run correctly, open the web browser, and show the index page.

Now we're at the end of this section, you should be able to understand the difference between an entity and a model and why it is important to not expose an entity to the client. You should be able to create models for your entity and create mapping profiles between them. You should also know how to create generic data services and how to register these services for the application.

Summary

After reading this chapter, you should know about EF Core and how to use it to create a code-first database in **MSSQL**, how to create migrations, and how to apply them to the server.

We also covered the topic of mapping the objects between each other using *AutoMapper* and why you should not expose the database schema to the client.

You should also have some idea about how to create generic services to manipulate the data.

By now, you should be able to create any kind of database entity with the corresponding business model returned to the client. Also, you should be able to create data services for CRUD operations.

In the next chapter, we will take a closer look at connecting the client Blazor WebAssembly application to our server part to consume the created data services.

Further reading

If you want to go deeper into some of the topics of this chapter, the following resources can provide more information:

- For more information on EF, refer to `https://docs.microsoft.com/en-us/ef/core/`

- For more information on C# generic classes and methods, refer to `https://docs.microsoft.com/en-us/dotnet/csharp/fundamentals/types/generics`

- For more information on *1:1*, *1:N*, and *M:N relations*, refer to the following links:

  ```
  https://database.guide/database-relationships-explained/
  https://en.wikipedia.org/wiki/One-to-one_(data_model)
  https://en.wikipedia.org/wiki/One-to-many_(data_model)
  https://en.wikipedia.org/wiki/Many-to-many_(data_model)
  ```

4

Connecting Client and Server with REST API

Having built the server side of our application, which can provide data for our Blazor WebAssembly application, we need to create some components that can consume the data. We can create these in a non-generic way and type a lot of code, or create generic components that will handle the same logic for each entity we have already created.

In this chapter, we will learn how to create Blazor components with generic parameters and how we can benefit from them in our application. We will also learn how to create the API endpoint to provide the data for our application.

In this chapter, we will cover the following topics:

- Exposing CRUD operations in API controllers
- Consuming a REST API in Blazor components

Technical requirements

All the code for this chapter can be found at `https://github.com/PacktPublishing/gRPC-Powered-Blazor-WebAssembly-Development/tree/main/ch4`.

Exposing CRUD operations in API controllers

The client side of our Blazor WebAssembly application will communicate with the server side using the HTTP client for API calls. We need to expose the endpoints for the client. Since we have a generic service `class`, we should create a generic controller.

We can start with the definition of the controller as follows:

Controllers\BaseController.cs

```
using Microsoft.AspNetCore.Mvc;
namespace MediaLibrary.Server.Controllers;


[ApiController]
[Route("rest/[controller]")]
public class BaseController<TModel, TEntity, TService> :
  ControllerBase
    where TModel : Shared.Model.IModel, new()
    where TEntity : Data.BaseEntity
    where TService : Services.BaseService<TEntity, TModel>
{
    private readonly TService _service;
    private readonly string _createPath;

    public BaseController(TService service, string
      createPath)
    {
        _service = service;
        _createPath = createPath;
    }


    // TODO: public methods
}
```

The preceding code shows the base class for our controllers. The BaseController class has three generic parameters. The first two, TModel and TEntity, are the same as in the BaseService class. The third is the service class itself.

Because we have registered the service classes in Program.cs, TService will be injected automatically for us. The createPath parameter has to be specified in the controllers and should contain the URL part to navigate to the newly created record in the client application.

For the create operation, we will use HTTP POST:

```
[HttpPost]
public virtual async Task<IActionResult> Create([FromBody]
```

```
    TModel model)
{

    model = await _service.CreateAsync(model);
    return Created($"{_createPath}/{model.Id}", new { Id =
      model.Id });
}
```

The preceding code shows the Create method, which will be available on the /rest/[controller] path for POST requests. This method will call the CreateAsync method on the service and return the path to the newly created item, with the item ID.

HTTP PUT will be used for the update:

```
[HttpPut("{id:int}")]
public virtual async Task<IActionResult> Update([FromRoute]
  int id, [FromBody] TModel model)
{

    await _service.UpdateAsync(id, model);
    return Ok();
}
```

The preceding code shows the Update method available on the /rest/[controller]/{id} path. The method simply calls the UpdateAsync method on the service.

For retrieving a single item, the HTTP GET request will be used:

```
[HttpGet("{id:int}")]
public virtual async Task<TModel> Get(int id)
{

    return await _service.GetByIdAsync(id);
}
```

The preceding code shows the Get method on the /rest/[controllers]/{id} path simply calling the GetByIdAsync method on the service and returning the data model.

When requesting the collection of data, the HTTP GET request is also used, but without the specified Id:

```
[HttpGet("list")]
public virtual async Task<IEnumerable<TModel>> GetList()
{

    var data = await _service.GetAllAsync();
```

```
        return data;
    }
```

The preceding code shows the `GetList` method available on the `/rest/[controllers]` path for GET requests. The method calls `GetAllAsync` on the service class. The returned data is a collection of the model class.

Finally, we would like to expose the endpoint for deleting the item. The path is the same as in the GET example, but instead of GET, the DELETE operation is used:

```
[HttpDelete("{id:int}")]
public virtual async Task<IActionResult> Delete(int id)
{
    await _service.DeleteAsync(id);
    return NoContent();
}
```

The preceding code shows the `Delete` method available on the `/rest/[controller]/{id}` path. The method calls `DeleteAsync` on the service class and returns a `NoContent` response.

Having implemented the `BaseController<TModel, TEntity, TService>` class, we should use this class for our `Movie` and `Person` entities.

The following example shows an implementation of `MovieController`:

Controllers\MovieController.cs

```
using MediaLibrary.Server.Services;
namespace MediaLibrary.Server.Controllers;

public class MovieController :
  BaseController<Shared.Model.MovieModel, Data.Movie,
    MovieService>
{
    public MovieController(MovieService service) :
      base(service, "/movies")
    {
    }
}
```

The preceding code shows the whole content of the file. The `MovieController` file is created, inheriting the `Base class Controller` and specifying the generic properties. The controller then passes the service class parameter to the base class and specifies the `createPath` parameter as `/movies`.

The same can be done for `PersonController`:

Controllers\PersonController.cs

```
using MediaLibrary.Server.Services;
namespace MediaLibrary.Server.Controllers;

public class PersonController :
  BaseController<Shared.Model.PersonModel, Data.Person,
    PersonService>
{
    public PersonController(PersonService service) :
      base(service, "/persons")
    {
    }
}
```

The preceding code shows the whole content of the file for `PersonController`. The logic here is the same as in `MovieController`.

Now, you should be able to create generic controllers and use them to expose endpoints for your client application. In the next section, we will take a look at how to use them.

Consuming a REST API in Blazor components

Up till now, we were focusing on the server side of the project. Now it is time to create some Blazor components and start using the code that we have already created. For our application, we will need two main components:

- **DataForm**: This component will take care of forms in our application and allow us to create and edit records

- **DataView**: This component will render a table to present existing data to the user

Creating and editing data

The Microsoft Blazor team has already created a lot of components that we can use. `InputText`, `InputDate`, `InputTextArea`, and many more form elements are already created for us. These form elements must be wrapped around the `EditForm` component to work properly.

The `EditForm` component will render the HTML `form` tag and provide the validation and submission logic to us. We don't want to write each detail (for the `Person` and `Movie` entities) from scratch, so we will prepare a single component that will handle all the logic. We will then use this component in our views and provide only the fields that we want to present to the user.

> **Note**
>
> All the files created or modified in this section will be located in the `MediaLibrary.Client` project.

Let's start with creating the `DataForm` Blazor component. In this component, we will be using the generic `TModel` parameter as our `Model` property. The generic parameter will then be marked as `CascadingTypeParameter` using the `@attribute` directive. This will allow us to use the `TModel` type in the child component without the need to specify the `TModel` type. The basic component should look as follows:

Shared\DataForm.razor

```razor
@typeparam TModel
@attribute [CascadingTypeParameter(nameof(TModel))]

<EditForm Model="Model" Context="FormEditContext">
    <DataAnnotationsValidator />
    <CascadingValue Value=Model>
        @ChildContent(Model)
    </CascadingValue>
    <hr />
    <button type="submit">Save</button>
</EditForm>
@code {
    [Parameter] public RenderFragment<TModel> ChildContent
      { get; set; }
    public TModel Model { get; set; } = new();
}
```

The preceding code shows the Blazor component for generic forms. The `EditForm` component provides the form logic, such as validating. The `DataAnnotationsValidator` component here tells `EditForm` that the `DataAnnotation` attributes from the model properties should be used in the validation. Then, the `CascadingValue` component passes the value of `Model` to all of the child components that have the parameter with the same `TModel` type and same name, `Model`.

> **Note**
>
> To use `DataAnnotation`, you can specify the attributes from the `System.ComponentModel.DataAnnotations` namespace to the `Model` properties. You can modify the `PersonModel` and `MovieModel` classes and annotate the `Name` property with the `[Required]` attribute. Add the `using` of the `System.ComponentModel.DataAnnotations` namespace to the top of the files to be able to use the `Required` attribute.

In the `@code` section of the file, we have defined the `ChildContent` and `Model` properties with the generic component type.

Because we will be writing more logic for this component, it is better to extract the `@code` block part to a separate file. Use the *Quick Action* icon (the bulb) or press *Ctrl + .* when the `@code` directive is selected to open **Quick Action** and select **Extract block to code behind**. This will create the `DataForm.razor.cs` file for us and move the definition of properties to that file. If you want, you can create the file manually.

The extracted *code-behind* file will contain many `using` directives that we don't need. After clearing the file of unnecessary lines, the file should look as follows:

Shared\DataForm.razor.cs

```
using Microsoft.AspNetCore.Components;
using System.Net.Http.Json;
using Microsoft.AspNetCore.Components.Forms;

namespace MediaLibrary.Client.Shared
{
    public partial class DataForm<TModel>
        where TModel : MediaLibrary.Shared.Model.IModel,
          new()
    {
        [Parameter]
        public RenderFragment<TModel> ChildContent
          { get; set; }
```

```
        public TModel Model { get; set; } = new();
    }
}
```

In the preceding code, you can see the base content of the DataForm base class. The difference from the generated file is in the following line:

```
where TModel : MediaLibrary.Shared.Model.IModel, new()
```

The preceding code shows the limitation of the TModel type in relation to the types that are implementing the IModel interface, and has an empty constructor.

Now, it is time to implement some more logic in the DataForm component. This component will use HTTP calls to the server to read and write data. Also, when the new record is created, we want to navigate the user to the correct URL address with the new record.

At first, we need to specify a few more parameters:

```
[Inject]
HttpClient Http { get; set; } = null!;
[Inject]
NavigationManager Navigation { get; set; } = null!;
```

The preceding code shows the two parameters in our component. The Inject attribute tells the component to use dependency injection and find the correct implementation of these types. The = null!; part in the code is to override the nullable validation check in the IDE, so we don't have any warnings.

We also need to know the path for the URL call and the ID of the record we want to modify:

```
[Parameter]
[EditorRequired]
public string ApiPath { get; set; } = string.Empty;


[Parameter]
[EditorRequired]
public int Id { get; set; }


private string _errorMessage = string.Empty;
```

The preceding code shows two parameters for our component. ApiPath will contain the path in the URL after /rest, and Id will contain zero for the new record or ID of the edited record. We are also defining the _errorMessage field, which will contain errors if any occur. The EditorRequired attribute tells the IDE to validate that these properties are specified when the component is used anywhere in the application.

We have all the information that we need in our components, so the next step is to use them. We can start with a method that will read the data from the server:

```
private async Task GetModel()
{
    Model = await
      Http.GetFromJsonAsync<TModel>($"rest/{ApiPath}/{Id}")
        ?? new();
}
```

In the preceding code, we are using the injected HttpClient in the Http property to call the URL on the server and parse the response to the TModel data stored in the Model property. If the response is null, the new TModel object is created.

We know that all our server endpoints have a rest/ prefix, the controller's name part, and Id if we want to manipulate a specific record. This unified approach allows us to create a simple universal call for data reading as shown in the preceding code.

The next step is to call this method when the component is rendered. To achieve this, we need to use the component OnInitializedAsync life cycle method:

```
protected override async Task OnInitializedAsync()
{
    await GetModel();
}
```

The preceding code shows the overriding of the component OnInitializedAsync life cycle method, where we are calling our defined GetModel method to retrieve data from the server.

The last method that we need is the method to save the form to the server. We can define the async SaveItem method and handle all the logic as follows:

```
private async Task SaveItem()
{
    HttpResponseMessage response = Id <= 0 ?
        await Http.PostAsJsonAsync($"rest/{ApiPath}",
          Model) :
```

```
        await Http.PutAsJsonAsync($"rest/{ApiPath}/{Id}",
          Model);

    if (response.IsSuccessStatusCode)
    {
        if (response.StatusCode ==
          System.Net.HttpStatusCode.Created)
        {
            if (response.Headers.TryGetValues("location",
              out var urls))
            {
                Navigation.NavigateTo(urls.First(),
                  replace: true);
            }
        }

        await GetModel();
    }
    else
    {
        _errorMessage = await
          response.Content.ReadAsStringAsync();
    }
}
```

In the preceding code, we have the SaveItem method, which handles the saving of data to the server. At the beginning of the method, we are creating the HttpResponseMessage object, which contains a response from the POST or PUT HTTP call depending on the ID parameter. For the new record, Id will be 0.

Next, we check the status code. If the response is a success, we also check whether the status code of the response indicates that the new item is created. If the item is new, we navigate the user to the new URL. We also reload data from the server because the server can modify our stored data. In case of an error, we are storing the error message in the _errorMessage field.

The last thing we need to do is to modify the `DataForm.razor` component file. We need to modify the `EditForm` component to use our `SaveItem` method by adding the `OnValidSubmit="SaveItem"` attribute:

```
// existing directives are here
<EditForm Model="Model" Context="FormEditContext"
  OnValidSubmit="SaveItem">
    // existing code is here
</EditForm>
```

The preceding code shows where to add the `OnValidSubmit` attribute. When the form is submitted, the `EditForm` component will validate the form and if successful, the `SaveItem` method will be called.

To present the error message, we can add the following code before the `EditForm` closing tag:

```
...
@if (!string.IsNullOrWhiteSpace(_errorMessage))
{
    <div class="alert alert-danger">@_errorMessage</div>
}
</EditForm>
```

The preceding code will render the error message when there is one

Now we have completed the form component and can test it on our first page. The page files are Blazor components. The only difference is that they have the @page directive at the top of the file:

Pages\PersonDetail.razor

```
@page "/persons/{Id:int}"
<DataForm ApiPath="person" Id="Id"
  TModel="MediaLibrary.Shared.Model.PersonModel"
  Context="model">
    <p><label>Name: <InputText @bind-Value="model.Name"
      /></label></p>
    <p><label>BirthDay: <InputDate @bind-
      Value="model.BirthDay" /></label></p>
    <p><label>BirthPlace: <InputText @bind-
      Value="model.BirthPlace" /></label></p>
```

```
<p><label>Biography: <InputTextArea @bind-
    Value="model.Biography" /></label></p></DataForm>

@code{
    [Parameter]
    public int Id { get; set; }
}
```

The preceding code shows the whole `PersonDetail` component. At the top of the file, we are using the @page directive to specify the page where this component lives. We are also using the Id parameter in the URL, so we are creating the property with the same name and type in the @code block.

This page uses the created `DataForm` component, saying `ApiPath` is a `person` and `Id` is from the URL route. The `Context` parameter on `DataForm` states the name of the property that we can use in the inner HTML. The inner HTML is then passed to the `ChildContent` parameter with the `RenderFragment<TModel>` type, so the IDE knows that the model is of the TModel type.

Now it's time to test our first page. You can start the project using *Ctrl + F5*, or from the menu, click on **Debug | Start Without Debugging**. The index page will be loaded in the browser.

Because we don't have any navigation yet, open the `https://localhost:7000/persons/0` URL in the browser. The form should be rendered. Filling in the form and clicking the **Save** button should save the data and navigate to the URL with the `Id` of the record.

> **Note**
> The port in the URL can be different depending on your local environment.

Now we can create new records and modify existing ones. However, we need to know the URL of the records first. Let's build the table view of existing data to be able to select the record to modify, or create a new record.

Viewing the data

When we want to render the data to the table, we have more options on how to use it. One of the common approaches is creating the component as a table wrapper and using the `ChildContent` parameter to render each row of the table. This allows us to modify the columns presented in the table, but forces us to specify the `ChildContent` parameter anywhere we are using the component.

In our demo, we will use a different approach. We will create a component that will load data from the given endpoint and render a column for each property in the returned data. To be able to present data as a table, we need to create some C# structures similar to the table:

Shared\Model\TableCell.cs

```
namespace MediaLibrary.Client.Shared.Model;
public class TableCell
{
    public object? Value { get; set; }
}
```

The preceding code shows the class for a single cell in the table. The cell can contain only the Value as we don't know the exact type we are using for the object.

We need to be able to present the whole table row:

Shared\Model\TableRow.cs

```
namespace MediaLibrary.Client.Shared.Model;
public class TableRow<TItem>
{
    public TableRow(TItem originValue)
    {
        OriginValue = originValue;
    }

    public List<TableCell> Values { get; set; } = new();
    public TItem OriginValue { get; set; }
}
```

The preceding code shows the class for a single table row. We are using the generic TItem property, which will hold the original value of the data, a single item in the collection, list, or array, and the list of values.

When we have the rows, we need to specify table columns to render the table header:

Shared\Model\TableColumn.cs

```
using System.Reflection;
namespace MediaLibrary.Client.Shared.Model;
public class TableColumn
{
    public string Name { get; set; } = string.Empty;
    public PropertyInfo PropertyInfo { get; set; } = null!;
}
```

In the preceding code, you can see the model of TableColumn. This class holds information about a single property. We are keeping the name of the column, which is the name of the property, and PropertyInfo to be able to check the type of the values in the column.

The last model is the Table class, which will have a collection of columns and rows:

Shared\Model\Table.cs

```
namespace MediaLibrary.Client.Shared.Model;
public class Table<TItem>
{
    public IEnumerable<TableColumn> Columns { get; set; } =
        new List<TableColumn>();
    public List<TableRow<TItem>> Rows { get; set; } =
        new();
}
```

The preceding code shows how the Columns and Rows properties are defined.

Let's combine all the code in the new component called DataView. This component will render the table with a column for each property in our model and a row for each record in our database. It will also show the navigation buttons to navigate the user to the detail of the item or to create a new item.

Start with creating two files, `DataView.razor` and `DataView.razor.cs`. We will first create the logic in the code behind the file, then we will create the HTML content of our component:

Shared\DataView.razor.cs

```
using Microsoft.AspNetCore.Components;
using System.Net.Http.Json;
using MediaLibrary.Client.Shared.Model;

namespace MediaLibrary.Client.Shared
{
    public partial class DataView<TItem>
        where TItem : MediaLibrary.Shared.Model.IModel,
          new()
    {
        // The code goes here
    }
}
```

The preceding code shows the basic structure of the `DataView` class. As in the `DataForm` component, we are creating a generic component with the same type as the `TItem` generic parameter.

Our component will be able to load the data by itself. To allow that, we need to inject the `HttpClient` class. We will also use the navigation, so the `NavigationManager` class needs to be injected as well. Other than the injected classes, we will need the `ApiPath` parameter and the property that will contain loaded data from the API. Add the following code to the component:

```
[Inject]
public NavigationManager Navigation { get; set; } = null!;
[Inject]
public HttpClient Http { get; set; } = null!;
[Parameter]
[EditorRequired]
public string ApiPath { get; set; } = string.Empty;
public Table<TItem> Data { get; set; } = new
  Table<TItem>();
```

The preceding code shows how to inject classes into our component and the definition of the public properties. The Data property will hold the table structure of the requested data.

Next, we need to call the API to download the data when the component is rendered on the page. Add the following method to the DataView component:

```
protected override async Task OnInitializedAsync()
{
    var type = typeof(TItem);
    Data.Columns = type.GetProperties().Select(x => new
        TableColumn { Name = x.Name, PropertyInfo = x });

    var model = await Http.GetFromJsonAsync
        <IEnumerable<TItem>>($"/rest/{ApiPath}/list") ?? new
        List<TItem>();

    foreach (var item in model)
    {
        var row = new TableRow<TItem>(item);

        foreach (var column in Data.Columns)
        {
            var value = column.PropertyInfo.GetValue(item);
            row.Values.Add(new TableCell { Value = value
                });
        }

        Data.Rows.Add(row);
    }
}
```

The preceding code will create the table columns from the type specified in the TItem parameter. Then, the API call to the server endpoint is made and the returned data is transformed into the table structure. All the data is stored in the Data property.

When creating the HTML part of our component, we will need to navigate the user to the detail. For this purpose, we can create the following method:

```
public string GetDetailUrl(int id)
    => $"{Navigation.ToAbsoluteUri(Navigation.Uri)
        .LocalPath}/{id}";
```

The preceding code shows the `GetDetailUrl` method, which will create a link to the current page followed by /{id}.

Now, we can create the HTML content for our component:

Shared\DataView.razor

```
@typeparam TItem
@using Microsoft.AspNetCore.Components.Web.Virtualization
<a href="@GetDetailUrl(0)" class="btn btn-
    secondary">New</a>
<table class="table table-bordered table-hover
    table-striped">
    <thead>
        <tr>
            @foreach (var column in Data.Columns)
            {
                <th>@column.Name</th>
            }
            <th></th>
        </tr>
    </thead>
    <tbody>
        @* The code goes here *@
    </tbody>
</table>
```

The preceding code shows the initial part of the file code, which will be extended later in this section. In the beginning, we specified the `type` parameter. Then, we used the `@using` directive to add the namespace for the components we will use later. The link to the detail with ID 0 is created and then the table.

The table contains a header with a `foreach` loop to render the column for each property in the model, and one more, which will be used to render the action controls.

To render the rows of the table, we can use the simple `for` or `foreach` loop, or use some advanced techniques:

```
<Virtualize Items="@Data.Rows" Context="item">
    <tr>
        @foreach (var data in item.Values)
        {
            <td>
                @if (data.Value is null)
                {
                    <span>---</span>
                }
                else
                {
                    @data.Value
                }
            </td>
        }

        <td class="is-actions-cell">
            <div class="buttons is-right">
                <a href="@GetDetailUrl
                    (item.OriginValue.Id)" class="btn
                    btn-primary">Edit</a>
            </div>
        </td>
    </tr>
</Virtualize>
```

The preceding code shows how to render the rows for our table. Instead of the standard loops available in Blazor, we are using the `Virtualize` component from the `Microsoft.AspNetCore.Components.Web.Virtualization` namespace. This component has two parameters: the `Items` parameter is used to specify the collection of data to render, and the `Context` parameter then tells the IDE what parameter should be available in the inner HTML.

The Virtualize component will then render one ChildContent for each record presented in the Items collection. The difference between the standard loop and the Virtualize component is that this component will create only the elements visible on the website viewport. This means that if you pass a collection with millions of records to the Items parameter, only a few of them will be rendered to the HTML. If you scroll the page, the elements get updated. This leads to massive performance improvement.

> **Note**
>
> If you want to limit the columns presented in the table, you can create a custom Attribute class and use this class on the properties you want to show in the table view. The component can then render only the columns for the properties with an attribute.

Now we have our DataView component and we can use it on the page:

Pages\PersonList.razor

```
@page "/persons"
<DataView ApiPath="person"
    TItem="MediaLibrary.Shared.Model.PersonModel" />
```

The preceding code shows all the necessary code to use the DataView component to render the Person records from our database.

To allow users to use this page, modify the NavMenu component. Add the following code to the nav element:

```
<div class="nav-item px-3">
    <NavLink class="nav-link" href="persons">
        <span class="oi oi-person"
            aria-hidden="true"></span> Persons
    </NavLink>
</div>
```

The preceding code shows the single link to the page. The NavLink component is used to specify the link location. The generated anchor will have an active class when the URL matches the href attribute.

Wow! We have created our first components, and these are not the easiest ones to create. So, well done if you have followed till now! To validate the demo project, run it without the debug mode. You should see the **Persons** link in the menu. When clicking on it, the table with columns for each PersonModel property should be rendered. Try adding new records and editing the existing ones.

Extending the demo project

We have implemented pages for the `Person` entity, but we also have to work on the `Movie` entity. We can create both pages for the `Movie` entity easily because we already have the components:

Pages\MovieList.razor

```
@page "/movies"
<DataView ApiPath="movie"
  TItem="MediaLibrary.Shared.Model.MovieModel" />
```

The preceding code shows the whole file for the table view of movies.

The `MovieDetail` page can be made easily using the `DataForm` component and the correct model properties:

Pages\MovieDetail.razor

```
@page "/movies/{Id:int}"
<DataForm ApiPath="movie" Id="Id"
  TModel="MediaLibrary.Shared.Model.MovieModel"
  Context="model">
    <p><label>Name: <InputText @bind-Value="model.Name"
      /></label></p>
    <p><label>Year: <InputNumber @bind-Value="model.Year"
      /></label></p>
    <p><label>Description: <InputText
      @bind-Value="model.Description" /></label></p>
    <p><label>Director Id: <InputNumber
      @bind-Value="model.DirectorId" /></label></p>
    <p><label>Music Composer Id: <InputNumber
      @bind-Value="model.MusicComposerId" /></label></p>
</DataForm>

@code {
    [Parameter]
    public int Id { get; set; }
}
```

The preceding code shows the whole file for editing the movies. Right now, we are using the IDs of persons specifying them as directors or music composers.

You can extend the NavMenu component again to contain a link to the movies page.

In this section, we have created two generic Blazor components that are reusable for each data model we want to create in the future, and we have learned how to create the components that will take care of the data itself.

Summary

After reading this chapter, you should know how to create an endpoint on the server side of the Blazor WebAssembly application. You should also know how to create these endpoints in a generic way and how to call the endpoints from the client side of the application.

We also covered the topic of creating the Blazor components with generic parameters and how to connect these components to consume the REST endpoints to get data to the client or send data to the server.

By now, you should be able to create your own Blazor components that can generate dynamic HTML content for websites. You should also know more about generic parameters in the components and how to inject the services on the client side of the Blazor WebAssembly application.

In the next chapter, we will take a closer look at gRPC services. We will introduce two ways of defining the gRPC services in C# applications and discuss the benefits and drawbacks of each. We will then create a copy of the existing DataView and DataForm components and modify them to use new gRPC services instead of HTTP API calls.

5

Building gRPC Services

By having a functional REST API implemented in our *MediaLibrary* application, we now have a fully functional website. While we can be satisfied with this state, we can also improve the performance of the application by switching from REST to gRPC services and bringing a more powerful communication protocol into our Blazor WebAssembly application.

In this chapter, we will learn how to create gRPC services in C# and how to consume them in the Blazor WebAssembly application. We will learn more about `.proto` files and present two ways of implementing gRPC in .NET applications. We will cover some advantages and disadvantages of both approaches.

Then, we will learn how to create Blazor components to consume gRPC services in a generic way.

By the end of this chapter, you will understand how gRPC works, the syntax of a Google Protocol Buffer language, and how to implement the services in C#. You will also learn how to consume these services from Blazor WebAssembly applications.

In this chapter, we will cover the following topics:

- What are the benefits of using gRPC services?
- How does communication work in gRPC?
- Understanding the protocol buffer language
- gRPC services in C#
- Creating Blazor components

Technical requirements

All the code for this chapter can be found at `https://github.com/PacktPublishing/gRPC-Powered-Blazor-WebAssembly-Development/tree/main/ch5`.

What are the benefits of using gRPC services?

gRPC is a powerful, language-neutral open source framework. Using TCP connection and binary serialization is more powerful and faster than the standard REST API communication with JSON serialization used in all JavaScript frameworks. gRPC also uses the *HTTP/2* protocol instead of the older *HTTP/1.1*.

The problem with gRPC is that it is a protocol primarily created for client-to-server communication, where the client is another server and both the client and server must support the HTTP/2 protocol. Browsers, on the other hand, do not support HTTP/2 yet.

This is where the Blazor comes in. Blazor, despite being the client WebAssembly part of the application, runs in the browser, supports the gRPC protocol, and allows us to use this powerful tool.

Here are the main benefits of using the gRPC protocol:

- **Binary serialization**: The protocol produces smaller messages than JSON, but it is not human-readable. Binary serialization converts the object into a stream of bytes. These bytes are then written to a data stream and sent to the client. The smaller size means a smaller amount of data is transferred over the network, which leads to faster loading times.

- **Protocol buffers**: gRPC has faster serialization using the protocol-buffer message format. Elements are serialized as key-value pairs. But because the client and the server know the position of the elements instead of the property name as a key, the order of property is used instead.

  We can show the difference in the following example:

  ```
  public class Person
  {
    public string FirstName { get; set; }
    public string MiddleName { get; set; }
    public string LastName { get; set; }
    public int Age { get; set; }
  }

  Person person = new Person {
    FirstName = "Andy",
    LastName = "Dufresene",
  }
  ```

  The preceding code shows the definition of the `Person` class and one instance of this class saved in the `person` variable.

The following code shows how the `person` properties will be serialized to the JSON format:

```
{
    "FirstName": "Andy",
    "MiddleName": "",
    "LastName": "Dufresene",
    "Age": 0
}
```

As you can see in the preceding code, the `MiddleName` and `Age` properties are serialized.

The following code shows how the `person` properties will be serialized to the gRPC format before the binary serialization:

```
1: "Andy"
2:
3: "Dufresene"
4:
```

In the preceding code, you can see pseudo-serialization of the gRPC protocol. It shows us how the data is presented in the protocol buffers. You can see that, instead of the property name, the order of property is used.

> **Note**
> The order of properties has to be defined in the `.proto` file.

The elements with a default value are also skipped in serialization. The default value for the client and the server should be the same, so there is no reason to transfer the value over the network.

- Support for **HTTP/2**: The HTTP/2 protocol brings us a new way of communication comprising binary communication using binary serialization and streaming, where one side of the communication can keep the connection to the other side open and send messages continuously. In HTTP/1.1, the client had to open the connection to the server each time. However, now, with HTTP/2, the server can keep the connection open and send the data continuously.

Now that you know the key benefits of using gRPC services, let us understand how communication works in gRPC. While REST uses standard request-response communication, gRPC supports multiple types of communication. We can look at them in the next section.

How does communication work in gRPC?

gRPC supports four types of communication: unary, client streaming, server streaming, and bi-directional streaming (sometimes called full-duplex). You can see the communication schema in the following diagram:

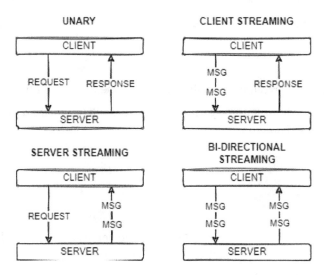

Figure 5.1 – A diagram of the gRPC communication types

Here are the definitions of each communication type:

- **Unary** communication is the standard form of communication in REST. It represents a request from the client and then a response from the server.

- **Client streaming** represents the client being able to send multiple messages to the server. When the client sends the last message, the server responds with a single response.

- **Server streaming** is the opposite of client streaming. The client sends a single request and the server responds with multiple messages.

- **Bi-directional streaming** is the combination of client and server streaming methods. The client and server are sending messages to each other at the same time.

Now, we know the benefits of using gRPC over a REST API and what kind of communication we can use in gRPC. In the next section, we will take a closer look at the implementation of gRPC in .NET and C#.

Understanding the protocol buffer language

The Google Protocol Buffer language (sometimes called protocol buffer language) is used to describe the structure of protocol buffer data. The language is currently in *version 3* and the `.proto` files should have directives at the top of the file specifying the version of the language used in the file (the current version is *proto3*). The file can also have a package name for sharing the files between applications.

The following code shows the basic directive definition of the file:

```
syntax = "proto3";
package grpc;
```

In the preceding code, we are using the `syntax` directive to specify the version and the `package` directive to specify the name of the package.

When we want to extend the package functionality, we can use the `option` and `import` directives:

```
option csharp_namespace = "MediaLibrary.Contracts";
import "different_file.proto";
```

In the preceding code, we are using the `option` directive to specify the C# namespace for generated code. We are also using an `import` directive to import types from another file.

Creating messages

Objects passed to the service methods and returned are called messages. The following example shows a simple definition of the object used to search:

```
syntax= = "proto3"
message SearchRequest {
   string search = 1;
   int32 maxResults = 2;
}
```

In the preceding example, you can see that the properties in the message have a specified numeric order. Because of this, the serialization can omit the property names when serializing data.

Field types

We are using a scalar type of the fields, but you can also use composite types, such as enumerations or other defined messages. The following scalar types are available in `.proto` files with their corresponding C# types:

Protobuf type	C# type
double	double
float	float
int32	int
int64	long
uint32	uint
uint64	ulong
sint32	int
sint64	long
fixed32	uint
fixed62	ulong
sfixed32	int
sfixed62	long
bool	bool
string	string
bytes	ByteString

Table 5.1 – Protobuf versus C# types

The difference between int, uint, sint, fixed, and sfixed (in both 32-bit and 64-bit versions) is in the encoding and size consumed in serialization. While int is better for standard use, sint is better when we expect mostly negative values. fixed can perform better when storing values greater than 2^{28}.

String values must contain UTF-8 encoded or 7-bit text. Both string and bytes cannot be longer than 2^{32}.

For a collection of items, the repeated keyword is used before the type definition:

```
repeated double Temperatures = 1;
```

The preceding code shows the property named Temperatures, which will be translated to the Google.Protobuf.Collections.RepeatedFiled<T> type, where T is the double type in our example.

Default values

The default value of the fields is the same as in C#. The only rule here is that for the Enum types, the default value is the first defined enum value, which must be 0.

The default value for `repeated` types is an empty collection. For fields with type as another message, the `null` value is a default in C#.

The reserved keyword

When defining the message or `enum`, you must specify the position of each field. There can be situations where for backward compatibility or future extensibility, you want to reserve some positions:

```
message Response {
    reserved 1, 3 to 5;
    string Message = 2;
    string Data = 6;
}
```

In the preceding code, we have used the `reserved` keyword to protect positions 1, 3, 4, and 5 in the message. This keyword can be used in certain scenarios, such as when in the past, the 1 position contained some field that we do not need anymore, or in the future, we would like to add some additional data to positions 3, 4, and 5.

Defining methods

Methods in the proto-language are represented by the `rpc` keyword. Each method must be in the wrapper, called `service` – equivalent to the C# `class`.

Methods are defined with the following schema:

```
rpc MethodName(RequestData) returns (ResponseData);
```

The preceding code shows that at the beginning, we need to specify the name of the method and then the type of input argument. At the end of the method, we are specifying the output of the method.

Each method must have an input parameter and an output parameter. If you do not need any data passed in or out, you can define a `message` without any field.

In the following example, you can see the full implementation of the `proto` file:

```
syntax = "proto3";
service WeatherService {
    rpc GetCurrentWeather(Location) returns
      (WeatherResponse);
}
message Location {
    string Address = 1;
```

```
}
message WeatherResponse {
  double Temperature = 1;
}
```

In the preceding code, we have defined two messages and one service with one method.

In this section, we have learned about the structure and syntax of the Google Protocol Buffer language. Now, we know how to create .proto files with the message and service definition. Next, we will look at how to use them in the C# .NET projects. Do we need the .proto files or not?

gRPC services in C#

Before we implement the gRPC service in our project, we need to understand how the implementation can be done in .NET projects in general.

Two ways of implementing

When we want to implement gRPC services in the .NET project, we have two options. The **Code-First** method, where we decorate the C# classes with attributes, and the standard way using .proto files. Let's start with the Code-First method.

Code-First

This method uses C# contracts: classes and interfaces in C# with attribute decoration. The advantage of this approach is that you do not need to learn any language other than C# and if you already have defined your models and interfaces, you can just decorate them instead of creating new models.

The main disadvantage, however, is that this approach is only useful if you plan to stay in the .NET technologies and you do not need to share the gRPC service definitions with other languages.

The Code-First method requires a NuGet package called protobuf-net.Grpc.

Now, it is time to look at the second option.

.proto files

Creating gRPC services using .proto files is a standard form of implementation. The .proto files can be shared across multiple platforms and used to generate services in different languages.

Each language supporting gRPC has a **protobuf compiler**: the mechanism that can read the .proto files and generate client implementations and server descriptions of the services and objects used in it.

The main difference between these two options is that using .proto files is a more universal method, but the Code-First method can be faster for development when sticking to the .NET platform.

Because we want to learn more about gRPC in general, we will use the `.proto` file option in the *MediaLibrary* project.

Implementing gRPC services using .proto files

To implement gRPC services, go through the following steps:

1. Open the **MediaLibrary** demo project and in **Solution Explorer**, right-click on the **Solution 'MediaLibrary'** item at the top of the window and choose **Add | New Project…**.

2. In the **Add a new project** window, find the **Class Library** project template in **Recent project templates**, or use the **Search for templates** input and click **Next**.

3. Set **Project name** to `MediaLibrary.Contracts` and click **Next**.

4. The **Additional information** window should be presented and the **.NET 6.0** version of the framework should be selected. If not, select the **.NET 6.0 (Long-term support)** version.

5. Click the **Create** button to create the project inside the current solution.

The new project should appear in **Solution Explorer**. This project only contains the `Class1.cs` file. We will not use this file, so it should be deleted.

Required packages

To use `.proto` files in our project, to bring the protobuf compiler to our project, and allow the C# code generation, respectively, we need to install the following packages to the `MediaLibrary.Contracts` project:

*   `Google.Protobuf` (*version 3.21.1*)
*   `Grpc.Net.Client` (*version 2.46.0*)
*   `Grpc.Tools` (*version 2.46.3*)

The fastest way to use these packages is to modify the `MediaLibrary.Contracts.csproj` file by double-clicking on it in **Solution Explorer** and adding the following lines before the `</Project>` closing tag:

```xml
<ItemGroup>
  <PackageReference Include="Google.Protobuf"
    Version="3.21.1" />
  <PackageReference Include="Grpc.Net.Client"
    Version="2.46.0" />
  <PackageReference Include="Grpc.Tools" Version="2.46.3">
    <PrivateAssets>all</PrivateAssets>
```

```
    <IncludeAssets>runtime; build; native; contentfiles;
        analyzers; buildtransitive</IncludeAssets>
  </PackageReference>
</ItemGroup>
```

The preceding code shows the package reference added to the project. Next, build the project by right-clicking on **MediaLibrary.Contracts** in **Solution Explorer** and clicking on **Build**. The referenced packages will automatically be downloaded and installed.

> **Note**
>
> You can also install the packages using *NuGet Package Manager* or the terminal (*CMD*, or *PowerShell*).

Creating proto files

First, we are going to create some shared types that we will use later in the services for manipulating persons and movies.

Start by adding a new shared.proto file:

1. Right-click on **MediaLibrary.Contracts** and select **Add | New Item**. Otherwise, hit *Ctrl + Shift + A* when the **Media.Library.Contracts** project is selected.

2. Type shared.proto into the **Name** field and click **Add**.

A new empty file should appear in **Solution Explorer**. Let's add some content:

shared.proto

```
syntax = "proto3";
package contracts.shared;
option csharp_namespace = "MediaLibrary.Contracts";

message ItemRequest {
  int32 id = 1;
}

message CreateResponse {
  int32 id = 1;
  string path = 2;
}
```

```
message GenericResponse {
    bool success = 1;
}


message Empty {}
```

The preceding code shows the whole shared.proto file. At the top of the file, we have directives specifying the language version of the file, the package name, and the C# namespace used for generated classes. Next, we have here a definition of four messages, which we will use in other .proto files.

Next, repeat the process to add another file called person.proto. In this file, we have to define the directives, import the types, and create custom messages and services that we want to be generated for us:

person.proto

```
syntax = "proto3";
package contracts;
option csharp_namespace = "MediaLibrary.Contracts";
import "google/protobuf/timestamp.proto";
import "shared.proto";
...
```

In the preceding code, you can see the basic definition is the same as in shared.proto. However, here, we have two import directives. The first imports the Timestamp type from the Google. Protobuf package. This type is translated to the C# DateTime. The second imports types from our shared file.

Next, we need to define the concrete message, which will represent our PersonModel class. We can define the message as follows:

```
...
message Person {
    int32 id = 1;
    string name = 2;
    google.protobuf.Timestamp birthDay = 3;
    string birthPlace = 4;
    string biography = 5;
    repeated int32 moviesIds = 6;
}
```

The preceding code shows the definition of the `Person` message. This message corresponds with the `PersonModel` class that we created before.

The last part is to create the service with the methods that we want to use. In our REST implementation, we are using CRUD methods. We need the equivalent here, so we need the `Create`, `Get`, `GetList`, `Update`, and `Delete` methods. We will put the following code between the top directives and the message definition:

```
...
service PersonContract {
    rpc Create(Person) returns
        (contracts.shared.CreateResponse);
    rpc Update(Person) returns
        (contracts.shared.GenericResponse);
    rpc Get(contracts.shared.ItemRequest) returns (Person);
    rpc GetList(contracts.shared.Empty) returns (stream
        Person);
    rpc Delete(contracts.shared.ItemRequest) returns
        (contracts.shared.GenericResponse);
}
...
```

The preceding code shows the definition of the `PersonContract` service with the required methods. For each method, we are specifying the input parameter and the return value. The `GetList` method returns a `stream` of `Person`, meaning the server can return multiple `Person` messages, but they will be returned from the server one by one. This will allow the client to process each `Person` individually.

Streaming data creates a lower requirement of memory allocation when transferring and parsing data. Also, the data can be presented to the client much faster, because the client did not need to wait for the whole collection of data.

Having the files in the project does not mean the code generation from the files begins. We need to modify `MediaLibrary.Contracts.csproj` once more and add the following lines before the `</Project>` element closing tag:

```
<ItemGroup>
  <Protobuf Include="shared.proto" />
  <Protobuf Include="person.proto" />
</ItemGroup>
```

These lines will notify the `Grpc.Tools` package to generate client and server implementations from our `.proto` files.

While `Grpc.Tools` generates the code for us, we need to create the exact server implementation of the service. The generated methods on the server are virtual methods returning `RpcException` with an `Unimplemented` status.

Server implementation of gRPC services

For the server part, we need to modify the *MediaLibrary.Server* project and add a reference to our *MediaLibrary.Contracts* project and also the `Grpc.AspNetCore` and `Grpc.AspNetCore.Web` packages:

1. Right-click on **Dependencies** in the **MediaLibrary.Server** project and select **Add Project Reference…**.

2. In the **Reference Manager – MediaLibrary.Server** pop-up window, check the checkbox next to **MediaLibrary.Contracts** and click **OK**.

3. Modify the `MediaLibrary.Server.csproj` file by double-clicking on the project name in **Solution Explorer** and adding the following lines to the **ItemGroup** section with the other **PackageReference** items:

    ```
    <PackageReference Include="Grpc.AspNetCore"
        Version="2.46.0" />
    <PackageReference Include="Grpc.AspNetCore.Web"
        Version="2.46.0" />
    ```

4. Close the file and build a project using *Ctrl + B* or go to the **Build | Build MediaLibrary.Server** option from the top menu.

Next, we need to tell our server project to use gRPC services. Let's modify the `Program.cs` file in the **MediaLibrary.Server** project by adding the following line of code to the **Add services to the container** section:

```
builder.Services.AddGrpc();
```

The preceding line of code tells our app to register gRPC services in the project initialization. This line has to be before the `builder.Build()` command is called.

The following line tells the app to use the gRPC services for the website and has to be added just after the `app.UseRouting();` command:

```
app.UseGrpcWeb();
```

Later, we will also register the concrete gRPC services to be available in the program. But first, we need to create them.

Create a directory called `Contracts` inside the **MediaLibrary.Server** project and then create a class file called `PersonContractService.cs` inside the `Contracts` directory.

Contracts\PersonContractService.cs

Remove the `using` directives at the top of the file and add the following `using` instead:

```
using AutoMapper;
using Grpc.Core;
using MediaLibrary.Contracts;
using MediaLibrary.Server.Services;
using MediaLibrary.Shared.Model;
...
```

The preceding code shows references to the required namespaces. `AutoMapper` will be used to map between the `Person` class created from the `proto` file and our `PersonModel` class. `Grpc.Core` contains types for gRPC service methods. `MediaLibrary.Contracts` points to our generated server implementation of the `PersonContract` service, and `MediaLibrary.Server.Services` and `MediaLibrary.Shared.Model` point to a namespace with our `PersonService` that we will use to manipulate data and the `PersonModel`.

Next, modify `PersonContractService` to inherit from `PersonContract.PersonContractBase` class. This class exists in the `MediaLibrary.Contracts` namespace and is generated for us by `Grpc.Tools`.

If we inspect the class by selecting `PersonContractBase` and hitting *F12*, we can see the generated **PersonGrpc.cs** file with `public abstract partial class PersonContractBase`. This class contains a `public virtual method` for each of our `rpc` methods defined in the `PersonContract` service. To implement these methods, we should override them in our `PersonContractService` class.

Close the `PersonGrpc.cs` file and navigate back to our `PersonContractService.cs` file. Inside the class, start by typing the `override` keyword. The context menu should appear and show us the methods that can be overridden. Select the **Create** method by clicking on it or by choosing it with arrow keys and hit *Enter*. Repeat this for the `Delete`, `Get`, `GetList`, and `Update` methods.

Notice that each of the methods has an additional `ServerCallContext context` parameter. This parameter can be used to access data about the caller or the cancellation token.

In the overridden methods, we will use AutoMapper to map types of properties in Person and PersonModel between each other and the existing PersonService to manipulate data. Let's add private fields and a constructor to the class:

```
private readonly PersonService _personService;
private readonly IMapper _mapper;

public PersonContractService(PersonService personService,
    IMapper mapper)
{
    _personService = personService;
    _mapper = mapper;
}
```

The preceding code shows the constructor of the PersonContractService class with two injected services, the PersonService and IMapper services.

Now, we can implement all methods. Start with the Create method:

```
public override async Task<CreateResponse> Create(Person
    request, ServerCallContext context)
{
    var model = _mapper.Map<PersonModel>(request);
    var created = await _personService.CreateAsync(model);
    return new CreateResponse { Id = created.Id,
        Path = $"/persons/{created.Id}" };
}
```

The preceding code shows the Create method. In the body of the method, we are using AutoMapper to map Person to PersonModel and then this model is passed to the PersonService method. As a return value, the CreateResponse class is used to return the created item identifier and the path to the created item.

Next, we can look at the Delete method:

```
public override async Task<GenericResponse>
    Delete(ItemRequest request, ServerCallContext context)
{
    await _personService.DeleteAsync(request.Id);
    return new GenericResponse { Success = true };
}
```

The preceding code shows that we are just passing the identifier from the request down to `PersonService`. Because the gRPC must return some value, we are returning the `GenericResponse` model.

The `Get` method uses `PersonService` to retrieve the record with the corresponding identifier and return it to the caller parsed to `Person`:

```
public override async Task<Person> Get(ItemRequest request,
    ServerCallContext context)
{
    var model = await
      _personService.GetByIdAsync(request.Id);
    return _mapper.Map<Person>(model);
}
```

The preceding code shows a simple implementation of the `Get` method.

`GetList` uses a stream as a return value, so it is a little different:

```
public override async Task GetList(Empty request,
    IServerStreamWriter<Person> responseStream,
    ServerCallContext context)
{
    var data = await _personService.GetAllAsync();
    foreach (var item in data)
    {
        var model = _mapper.Map<Person>(item);
        await responseStream.WriteAsync(model);
    }
}
```

In the preceding code, you can see that the return value of the `GetList` method is `Task`. Instead of returning the `stream` of `Person` from the method, the `IServerStreamWriter<Person>` parameter is passed to the method. The response data are then written to this stream one by one.

The last overridden method is `Update`:

```
public override async Task<GenericResponse> Update(Person
    request, ServerCallContext context)
{
    var model = _mapper.Map<PersonModel>(request);
```

```
        await _personService.UpdateAsync(request.Id, model);
        return new GenericResponse { Success = true };
}
```

In the preceding code, we are mapping `Person` to `PersonModel`, calling the `UpdateAsync` method on `PersonService`, and returning `GenericResponse` to the client.

Because we want to use the mapping between the generated `Person` class and our `PersonModel`, we need to register these classes for `AutoMapper`. The `Person` class also uses the `Timestamp` type, while `PersonModel` has `DateTime`. We need to map these types between each other as well. Because we will also use the mapping between these types in the client part of the application, let us create this in the `MediaLibrary.Shared` project.

Add a reference to the `AutoMapper` and `Google.Protobuf` NuGet packages and `MediaLibrary.Contracts` project to the `MediaLibrary.Shared` project.

Next, create a `SharedMapperProfile.cs` file and add the following using directive:

```
using AutoMapper;
using Google.Protobuf.WellKnownTypes;
```

The preceding code will allow us to use the `Timestamp` type in the mappings.

Next, add the inheritance of the `Profile` class and create the mapping by adding the following constructor code:

```
public SharedMapperProfile()
{
    CreateMap<DateTime, Timestamp>()
        .ConvertUsing(x =>
            Timestamp.FromDateTime(DateTime.SpecifyKind(x,
            DateTimeKind.Utc)));

    CreateMap<Timestamp, DateTime>()
        .ConvertUsing(x => x.ToDateTime());

    CreateMap<Contracts.Person, Model.PersonModel>()
        .ReverseMap();
}
```

In the preceding code, we are creating three maps. The first is between `DateTime` and `Timestamp`, where we are using the `FromDateTime` method and specifying the UTC format. The second map shows the mapping in the opposite direction of these types, by using the internal `ToDateTime` method on the `Timestamp` type. The last creates a two-way mapping between the `Person` and `PersonModel` classes.

It is time to navigate back to the `MediaLibrary.Server\Program.cs` file and add the final line to register our created service. Add the following line of code after the `app.MapControllers();` line:

```
app.MapGrpcService<MediaLibrary.Server.Contracts
    .PersonContractService>()
        .EnableGrpcWeb();
```

In the preceding code, the concrete service is mapped as a gRPC service and is marked to be available from the browser calls. Without the `EnableGrpcWeb` call, the request to the gRPC service from the browser will return a response, **415 Unsupported Media Type**, with a `grpc-message`, **Content-Type 'application/grpc-web' is not supported.**

The preceding line of code has to be added for each gRPC service that we create in the future but modified to use the concrete service class.

Also, extend the line with the `AddAutoMapper` command to use the newly created `SharedMapperProfile`. The result should look as follows:

```
builder.Services.AddAutoMapper(typeof(MapperProfile),
    typeof(MediaLibrary.Shared.SharedMapperProfile));
```

Having this, we need to modify the client application to be able to consume the gRPC services.

A client implementation of gRPC services

For the client implementation, we will need to reference the `MediaLibrary.Contracts` project, and the `Grpc.Net.Client` and `Grpc.Net.Client.Web` NuGet packages.

To create a reference to `MediaLibrary.Contracts`, follow the same steps as in the server implementation but select **Dependencies** inside the **MediaLibrary.Client** project.

To add the NuGet packages, modify `MediaLibrary.Client.csproj` and add the following lines of code after the other `PackageReference` lines:

```
<PackageReference Include="AutoMapper" Version="11.0.1" />
<PackageReference Include="AutoMapper.Extensions.Microsoft
    .DependencyInjection" Version="11.0.0" />
<PackageReference Include="Grpc.Net.Client.Web"
    Version="2.46.0" />
```

```
<PackageReference Include="Grpc.Net.Client"
    Version="2.46.0" />
```

Next, modify the `Program.cs` file inside **MediaLibrary.Client** and add the following `using` statements:

```
using Grpc.Net.Client;
using Grpc.Net.Client.Web;
using MediaLibrary.Contracts;
using Microsoft.AspNetCore.Components;
```

The `using` statements in the preceding code will allow us to consume generated gRPC clients and also use types from the `Grpc.Net.Client` namespace.

Lastly, add the following code before the `await builder.Build().RunAsync();` command:

```
builder.Services.AddAutoMapper(typeof(MediaLibrary.Shared
    .SharedMapperProfile));
builder.Services.AddSingleton(s =>
{
    var httpClient = new HttpClient(new
        GrpcWebHandler(GrpcWebMode.GrpcWeb, new
        HttpClientHandler()));
    var baseUri =
        s.GetRequiredService<NavigationManager>().BaseUri;
    var channel = GrpcChannel.ForAddress(baseUri, new
        GrpcChannelOptions { HttpClient = httpClient });
    var client = new
        PersonContract.PersonContractClient(channel);
    return client;
});
```

In the preceding code, we are registering the `AutoMapper` service and also creating a singleton instance of `PersonContractClient`. This client uses `HttpClient` internally configured to communicate with our server.

Each time we create a new gRPC service, we need to register a new singleton instance of the created client class to be able to consume the data from the server.

In this section, we have learned how to create the gRPC services in the C# projects and how to implement them to be available from the browser. We also now understand that there are two options for how to implement the gRPC and we know the advantages and disadvantages of each.

Wow! We have come quite a long way. Now, it is time to consume our first gRPC service in the Blazor WebAssembly application.

Creating Blazor components

In this section, we will take a closer look at how to consume the gRPC services in the Blazor WebAssembly application. But we have not yet tested that our implementation is working.

Start by simply building the solution using *Ctrl + Shift + B*. The solution should be built without any errors. If you see errors in your program, check that you follow all the steps correctly, or download the demo project from the GitHub folder (`ch5\01_demo_implementation`) to start with clean code.

First, we should test that our gRPC service is working correctly. Open the `_Imports.razor` file in **MediaLibrary.Client** and add `using AutoMapper` to the end of the file.

Next, open the `Pages\PersonList.razor` file and add the following directives to the top of the file, after the `@page` directive:

```
@using MediaLibrary.Shared.Model
@inject IMapper mapper
@inject MediaLibrary.Contracts.PersonContract
   .PersonContractClient client
```

The preceding code imports a namespace with the `PersonModel` class and also injects two services, `IMapper` from `AutoMapper`, and `PersonContractClient`, generated from the `.proto` file.

At the end of the file, add the `@code` block directive with the following content:

```
public List<PersonModel> Data { get; set; } = new
   List<PersonModel>();

protected override async Task OnInitializedAsync()
{
    var stream = client.GetList(new
      Contracts.Empty()).ResponseStream;

    while (await stream.MoveNext(default))
    {
        var item = mapper.Map<PersonModel>(stream.Current);
        Data.Add(item);
```

```
        }
    }
```

The preceding code defines the `Data` property, which is a list of `PersonModel`. The `Data` property is then filled in by the overridden `OnInitializedAsync` method from the generated client.

As you can see, the data is not returned as a collection but as a stream, which can be read one by one. Each item in the response is then mapped to `PersonModel` and attached to the `Data` property.

Now, we can comment out the `DataView` element used on this page, and add the HTML table to render the contents of the `Data` property:

```
<table class="table table-bordered table-striped">
    @foreach (var item in Data)
    {
        <tr>
            <td>@item.Id</td>
            <td>@item.Name</td>
            <td>@item.BirthDay</td>
            <td>@item.BirthPlace</td>
        </tr>
    }
</table>
```

The preceding code will render a table with one row for each record returned from the `PersonContractClient` service.

Now, make sure the *MediaLibrary.Server* project is set as the startup project. The project name in **Solution Explorer** will be in bold text. If not, right-click on the **MediaLibrary.Server** project and select **Set as Startup Project**. Run the project, and when the browser opens, select the **Persons** option from the left-hand menu. The table with persons will be rendered in the browser.

Now that we have validated that our gRPC service is working correctly, it is time to create a component that will use this service. But we don't want to only use this service, because we will create a similar service for *movies* later, and we want our component to work with it too.

To be able to create a Blazor control with a generic service, we need to have an interface describing the methods available in the client generated from the `.proto` file.

Preparation for generic components

Create an empty interface, `IContractClient.cs`, in the `MediaLibrary.Contracts\`
`Clients` directory and modify the namespace of the file to `MediaLibrary.Contracts`.
Also, add the `TItem` generic parameter to the `interface`, with the condition to limit `TItem` to
`class` with an empty constructor:

```
public interface IContractClient<TItem>
    where TItem : class, new()
{ ... }
```

Next, create the `PersonContract.cs` class file in the same directory and modify the namespace
as well. The name of the class will be underlined with a red line. When you hover the mouse over
the name, the **CS0260** error description is shown: **Missing partial modifier on declaration of type
'PersonContract'; another partial declaration of this type exists**. The reason for this error message
is that the `PersonContract` class is generated as `partial` in the `PersonGrpc.cs` file
generated from the `.proto` file.

Fixing the class using the `partial` keyword will remove the error message. The `PersonContract`
class contains a `partial PersonContractClient` class, and that is the client we will be
using to retrieve the data:

MediaLibrary.Contracts\Clients\PersonContract.cs

```
namespace MediaLibrary.Contracts;
public partial class PersonContract
{
    public partial class PersonContractClient :
      IContractClient<Person> { }
}
```

In the preceding code, we are creating another partial declaration of `PersonContractClient`
saying that this class has to implement our `IContractClient<TItem>` interface.

Our interface is currently empty. We need to specify the methods that we want to expose in the Blazor
component. To see what methods are implemented, click on the `PersonContractClient` class
name and hit *F12*. From the open window, select the implementation in **PersonGrpc.cs** by clicking
on it. The generated file is opened and we can inspect all the available methods.

You can see that the protobuf compiler generates four versions of each method defined in the `.proto` file.
The synchronous and asynchronous version is generated, both in the version with the `CallOptions`
parameter, or with the `Metadata`, `DateTime deadline`, and `CancellationToken`
optional parameters. The method with multiple parameters will internally call the method with

CallOptions, because the CallOptions parameter holds Metadata, deadline, and CancellationToken as well. Only the GetList method is not generated in the asynchronous version because the response of this method is a stream, not the exact value.

For our interface, we will use asynchronous versions of the methods, if available, in the version with the optional parameters:

MediaLibrary.Contracts\Clients\IContractClient.cs

```
using Grpc.Core;
namespace MediaLibrary.Contracts;


public interface IContractClient<TItem>
    where TItem : class, new()
{

    AsyncUnaryCall<CreateResponse> CreateAsync(Titem
      request, Metadata? headers = null, DateTime? Deadline
      = null, CancellationToken cancellationToken
      = default);
    AsyncUnaryCall<GenericResponse> UpdateAsync(Titem
      request, Metadata? headers = null, DateTime? Deadline
      = null, CancellationToken cancellationToken
      - default);
    AsyncUnaryCall<TItem> GetAsync(ItemRequest request,
      Metadata? headers = null, DateTime? deadline = null,
      CancellationToken cancellationToken = default);
    AsyncServerStreamingCall<TItem> GetList(Empty request,
      Metadata? headers = null, DateTime? deadline = null,
      CancellationToken cancellationToken = default);
    AsyncUnaryCall<GenericResponse> DeleteAsync(ItemRequest
      request, Metadata? headers = null, DateTime? Deadline
      = null, CancellationToken cancellationToken
      = default);
}
```

In the preceding code, you can see the declaration of all methods that we want to have available in the Blazor component. We have used Create, Update, Get, and Delete in asynchronous versions and GetList in synchronous versions.

Now, it is time to use the created interface in the generic Blazor component.

Creating generic list components

In *Chapter 4, Connecting Client and Server with REST API*, we created the `DataView` and `DataForm` components. Now, we will be creating modified versions of these components, beginning with the `DataView` component.

Navigate to `MediaLibrary.Client\Shared` and create a new Blazor component `GrpcDataView.razor` file.

Open the `DataView.razor` file and copy all the contents of the file. Paste the copied text into the `GrpcDataView.razor` file. To the top of the file, after the `@typeparam` directive, add the other `@typeparam` directives:

```
@typeparam TContractItem
@typeparam TContractClient
```

The preceding code defines two other generic parameters for the `GrpcDataView` component. The same parameters must be defined in the code behind, so create the `GrpcDataView.razor.cs` file and modify it.

At first, add the `using` of the `AutoMapper` namespace to the top of the file. Then, modify the class declaration to reflect the generic parameters created:

```
...
public partial class GrpcDataView<TItem, TContractItem,
    TContractClient>
        where TItem : class, MediaLibrary.Shared.Model.IModel,
            new()
        where TContractItem : class, new()
        where TContractClient : ClientBase<TContractClient>,
            IContractClient<TContractItem>
{
...
```

The preceding code shows the definition of the partial `GrpcDataView` class. The three generic parameters are defined and the conditions for the parameters are specified to meet the requirements of the types used. If any namespace is missing, add it using refactoring or **Quick Action**.

Now that we know that the client must inherit from `ClientBase` and implement `IContractClient`, we can create a property for this client. Add the following code inside the class declaration:

```
[Inject]
public TContractClient Service { get; set; } = null!;
```

The preceding code will tell Blazor to inject a service of the `TContractClient` type into the `Service` property when the component is attached to the page. We can also inject `AutoMapper` by adding another property:

```
[Inject]
public IMapper Mapper { get; set; } = null!;
```

The preceding code will create a `Mapper` property with an injected instance of `AutoMapper`. We will use `Mapper` after loading data from `Service`.

The `Http` and `ApiPath` properties can be removed from the `GrpcDataView` class because we will not need them.

In the `OnInitializedAsync` method, we will replace the line where the `Http` client was used to load data from the REST API with the following:

```
var stream = Service.GetList(new Empty()).ResponseStream;
```

The preceding line of code will call the injected `Service` class – the gRPC client – and return a response stream.

Next, we can replace `foreach` with the `while` loop, reading one item in `stream` at a time:

```
while (await stream.MoveNext(default))
{
    var item = Mapper.Map<TItem>(stream.Current);
// The rest of the loop stays the same as in foreach.
}
```

The preceding code shows how to read data from the `stream` response. The `MoveNext` method is called and awaited for every single record returned in the `stream` response. When there is no data, the loop is terminated.

Inside the loop, we are also using `Mapper` to map between the returned `TContractItem` and `TItem` models. In the case of persons, `TItem` will represent the `PersonModel` class from `MediaLibrary.Shared.Model`, and `TContractItem` and the `Person` class from the `MediaLibrary.Contracts` namespace.

Let's navigate back to the `PersonList.razor` component. Now, we can clean the component to the original state:

MediaLibrary.Client\Pages\PersonList.razor

```
@page "/persons"
<DataView ApiPath="person"
  TItem="MediaLibrary.Shared.Model.PersonModel" />
```

The preceding code shows the original state of the component at the beginning of this chapter. Now, we will replace the `DataView` component with the `GrpcDataView` component:

```
@page "/persons"
@using MediaLibrary.Contracts
@using MediaLibrary.Shared.Model
<GrpcDataView
    TItem="PersonModel"
    TContractItem="Person"
    TContractClient="PersonContract.PersonContractClient" />
```

In the preceding code, we have added `@using` directives to simplify the definition of the generic parameters. Then, we replaced the component and specified three generic parameters. `TItem` stands for the rendered model. `TContractItem` stands for the model handled in the gRPC client, and `TContractClient` stands for the type of the gRPC client itself.

Now, it's time to see the component in action. Run the project using *Ctrl + F5*. In the browser, open the developer tools by hitting *F12* and opening the **Network** tab. Now navigate to the **Persons** page using the left-hand menu.

In the **Network** tab, the POST call named **GetList** will appear. Inspecting the requested URL, we can see that the `https://localhost:7000/contracts.PersonContract/GetList` endpoint was called over the HTTP/2 protocol. If we want to read the response, *Firefox* will show us encoded data, while *Chrome* or *Edge* will try to decode the response (or at least parts of it) to the readable format.

We can now list the data using the gRPC service. But what about creating and editing data?

Creating generic form components

We can follow the same steps for the GrpcDataForm component as were followed for GrpcDataView. Start with creating the GrpcDataForm.razor and GrpcDataForm.razor.cs files. Then, copy the content of DataForm files into these newly created files and rename partial class in the GrpcDataForm.razor.cs file to correspond with a file name.

Then, add the @typeparam using directives to the top of the GrpcDataForm.razor file, just after the existing @typeparam:

```
@typeparam TContractItem
@typeparam TContractClient
```

The preceding code will allow us to use the generic parameters in the component.

Now, modify the GrpcDataForm.razor.cs file. Start with the class definition:

```
public partial class GrpcDataForm<TModel, TContractItem,
    TContractClient>
        where TModel : MediaLibrary.Shared.Model.IModel, new()
        where TContractItem : class, new()
        where TContractClient : ClientBase<TContractClient>,
            IContractClient<TContractItem>
```

The preceding code will modify the class definition to use the generic parameters.

In the same way as with the GrpcDataView component, remove the ApiPath and Http properties, and use the Service and Mapper properties instead:

```
[Inject]
public TContractClient Service { get; set; } = null!;

[Inject]
public IMapper Mapper { get; set; } = null!;
```

The preceding code will add the required functionality to the component.

Now, replace the body of the GetModel method with the following:

```
var grpcItem = await Service.GetAsync(new ItemRequest { Id
    = Id }) ?? new();
Model = Mapper.Map<TModel>(grpcItem);
```

In the preceding code, we are downloading the item with the passed `Id` from the gRPC client and then mapping it to `Model`.

The last thing to modify in the `GrpcDataForm` component is the `SaveItem` method:

```
var data = Mapper.Map<TContractItem>(Model);

if (Id <= 0)
{
    var response = await Service.CreateAsync(data);

    if (response.Id > 0)
    {
        Navigation.NavigateTo(response.Path, replace:
            true);
    }
}
else
{
    await Service.UpdateAsync(data);
}
```

The preceding code shows how to create or update data using the injected gRPC client in the `Service` property. If the new item is created, `Navigation` will navigate us to the URL address representing the detail of the newly created item.

To use `GrpcDataForm`, update the content of the `PersonDetail.razor` component. Add `using` to the two namespaces at the top of the file:

```
@using MediaLibrary.Contracts
@using MediaLibrary.Shared.Model
```

And replace the `DataForm` component with `GrpcDataForm`:

```
<GrpcDataForm Id="Id" TModel="PersonModel"
    TContractItem="Person"
    TContractClient="PersonContract.PersonContractClient"
    Context="model">
        … The inside of the component stays as it is.
</GrpcDataForm>
```

We have successfully created two Blazor components that can consume the gRPC services. They do not just consume them; they can create the instance of the service from the provided type parameters. We can now also repeat the process for the `Movie` entity:

MediaLibrary.Contracts\movie.proto

```
syntax = "proto3";
package contracts;
option csharp_namespace = "MediaLibrary.Contracts";
import "google/protobuf/timestamp.proto";
import "google/protobuf/wrappers.proto";
import "shared.proto";

service MovieContract {
  rpc Create(Movie) returns
    (contracts.shared.CreateResponse);
  rpc Update(Movie) returns
    (contracts.shared.GenericResponse);
  rpc Get(contracts.shared.ItemRequest) returns (Movie);
  rpc GetList(contracts.shared.Empty) returns
    (stream Movie);
  rpc Delete(contracts.shared.ItemRequest) returns
    (contracts.shared.GenericResponse);
}

message Movie {
  int32 id = 1;
  string name = 2;
  repeated CategoryType categories = 3;
  int32 year = 4;
  string description = 5;
  google.protobuf.Int32Value directorId = 6;
  google.protobuf.Int32Value musicComposerId = 7;
  repeated int32 actorIds = 8;
}

enum CategoryType {
```

```
Action = 0;
  Comedy = 1;
  Drama = 2;
  Fantasy = 3;
  SciFi = 4;
  Horror = 5;
  Mystery = 6;
  Romance = 7;
  Thriller = 8;
  Western = 9;
}
```

The preceding code shows the definition of `MovieContract`. You can see that we are importing the `wrapper.proto` file from `google/protobuf`. This file contains nullable types, such as nullable `int` and nullable `double` types.

Next, register the file in `MediaLibrary.Contracts.csproj` by adding the following line next to other `Protobuf` imports:

```
<Protobuf Include="movie.proto" />
```

Create the new file, `MovieContract.cs`:

MediaLibrary.Contracts\Clients\MovieContract.cs

```
namespace MediaLibrary.Contracts;
public partial class MovieContract
{
    public partial class MovieContractClient :
      IContractClient<Movie>
    { }
}
```

The preceding code defines the created `MovieContractClient` and implements the `IContractClient` interface.

Next, add the new mapping definitions to `SharedMapperProfile.cs`:

```
CreateMap<Contracts.Movie, Model.MovieModel>().ReverseMap();
```

The preceding code will create mappings between the existing MovieModel class and the generated Movie class.

Extend the Program.cs file in the MediaLibrary.Client project by adding a new singleton client:

```
builder.Services.AddSingleton(s =>
{
    var httpClient = new HttpClient(new
      GrpcWebHandler(GrpcWebMode.GrpcWeb, new
      HttpClientHandler()));
    var baseUri =
      s.GetRequiredService<NavigationManager>().BaseUri;
    var channel = GrpcChannel.ForAddress(baseUri, new
      GrpcChannelOptions { HttpClient = httpClient });
    var client = new
      MovieContract.MovieContractClient(channel);
    return client;
});
```

The preceding code will register a new client for communicating with gRPC services for movies.

Create a new file, MovieContractService.cs, in MediaLibrary.Server\Contracts. Add the inheritance of the MovieContract.MovieContractBase class and mark the class to implement the IContractService<Movie> interface.

Override the Create, Delete, Get, GetList, and Update methods the same way as we did in PersonContractService.

Extend the Program.cs file in the MediaLibrary.Server project. Add the following line of code after the registration of PersonContractService:

```
app.MapGrpcService<MediaLibrary.Server.Contracts
  .MovieContractService>()
    .EnableGrpcWeb();
```

Now, you can update the MovieList and MovieDetail components.

In this section, we have created two generic Blazor components that use the generated gRPC clients to communicate with the server side of the application.

Summary

After reading this chapter, you should know the syntax of the Google Protocol Buffer language, version 3. You should also know how to define messages and services in this language.

We also covered multiple options for implementing the gRPC services in the C# projects. We used the standard way of implementing the services using `.proto` files.

By now, you should be able to create gRPC services and consume them in the Blazor components. You should also know more about the nullable types in the gRPC and how to tell the protobuf compiler to generate client and server implementation from the `.proto` files.

In the next chapter, we will take a closer look at the source generators. We will define the classes that can be automatically generated for us and which we better type for ourselves.

6

Diving Deep into Source Generators

We started our journey in this book with an empty folder, within which we went on to create a Blazor WebAssembly application. Then, we added generic services to manipulate data and generic controllers to provide a REST API. Then, in *Chapter 5, Building gRPC Services*, we integrated the gRPC services that enable our Blazor WebAssembly application to communicate faster between the client and server parts of the application. So, it would not be wrong to say that we have built a functional application. And while it is true, there is always something that we can do better. In this chapter, we will take a look at an option that may speed up our development process: source generators.

As the name implies, source generators can generate source code for our application. In this chapter, we will learn what we can and can't do with source generators, how to write our generators, and how to consume the generated code. We will also take a look at the `partial` keyword, which is commonly used in the generated code.

At the end of this chapter, our application will benefit from source generators and provide an easy way to add functionality without us typing hundreds of lines of code. Our goal is to have a real-world Blazor WebAssembly application using gRPC to communicate between the client and server parts and with services generated with source generators.

We are almost at the end of our great journey of building a Blazor WebAssembly application with gRPC, and we are going to draw out our ending just a bit to add some important finishing touches.

In this chapter, we will cover the following topics:

- What can source generators generate?
- Using partial classes and methods
- Creating attributes for your needs

- Writing the first generator
- Exploring generated code

Technical requirements

All the code for this chapter can be found at `https://github.com/PacktPublishing/gRPC-Powered-Blazor-WebAssembly-Development/tree/main/ch6`.

What can source generators generate?

Can we use source generators to generate an `interface`? Yes, we can. And what about a `class`? Also yes. And a `method`? That's a third yes. Source generators can generate any kind of C# code.

This may sound like some kind of magic to you now, but I assure you that you still need to write a lot of code by yourself. So, don't worry about losing your job to some source generators.

As we mentioned in *Chapter 1, Introducing Blazor, gRPC, and Source Generators*, source generators can't modify existing code. When you're thinking about creating some source generators, keep this in mind.

On the other hand, they can create any code you want, including interface implementations, partial classes, enums, and more. Source generators can also read the input file (JSON, CSV, or any other format you can read in C#) and use the values from this input to generate the code.

Now, what is the benefit of using source generators? More than you can count! To mention a few, with source generators, you can generate the following:

- Interfaces from existing implementations
- Implementations from existing interfaces
- Repetitive code, such as an `INotifyPropertyChanged` implementation for properties
- Validators for `enum` values
- Anything you are using runtime reflection for

> **Note**
> Reflection allows us to dynamically create an instance of a type, but also get the type from an existing object. We can use reflection to read attributes on properties and classes during runtime.

While the performance of reflection is getting better and better in every version of .NET, it is still not as performant as typed code.

A lot of .NET applications use reflection in the startup process to register all the service implementations. With source generators, you can generate a method that will find all the services and generate some C# code to register all of them during design time.

Imagine having the IMovieProvider interface, which is implemented by services providing methods to manipulate the movies. To mention a few, we will have NetflixMovieProvider, DisneyMovieProvider, and HboMovieProvider. To register all of them, you can write the following code in your Configuration method in the Startup class:

```
services.AddTransient<IMovieProvider,
   NetflixMovieProvider>();
services.AddTransient<IMovieProvider,
   DisneyMovieProvider>();
services.AddTransient<IMovieProvider, HboMovieProvider>();
```

The preceding code shows a standard way of registering services to your **dependency injection (DI) container**. It is just three lines of code. But imagine a situation where you have 50 of them. You will probably change your code to something like this:

```
Assembly assembly = typeof(MediaLibrary.Startup).Assembly;
List<Type> types = assembly.GetTypes().Where(type =>
   type.GetInterfaces()
   .Contains(typeof(IMovieProvider))).ToList();
foreach (Type type in types)
{
   services.AddTransient(typeof(IMovieProvider), type);
}
```

In the preceding code, you can see how to use reflection to discover all the types that implement the IMovieProvider interface and how to register them for the DI container.

Because of reflection, the code here performs worse than the code that registers each service. The advantage of this approach is that we do not need to register the service class because it is done automatically.

Source generators can take advantage of both approaches and put them together. With the source generator, you can discover all the classes that implement the IMovieProvider interface during the build process and generate a method containing the service registration. Then, you only need to use the generated method to register the services:

```
// AutoGenerated file
public static void RegisterGeneratedMovieProviders(this
```

```
    IServiceCollection services)
    {
        // Generated code here, one line for each implementation
        of IMovieProvider.
    }
    // Program.cs file
    services.RegisterGeneratedMovieProviders();
```

The preceding code shows an example of a generated method and the usage of this method in the Program.cs file. Because we don't use reflection here, the performance is better. Also, we don't need to take care of every implementation because source generators can find the implementations when exploring the code during the compilation process.

This was just an example of when you can use source generators instead of reflection. In a real-world application, reflection during the startup process is not something you care about so much, because it is a one-time operation. Instead of doing this, you can create a generator, which will replace the mappings between different types in your code, or other operations, which use reflection and are executed many times during the life of the application. If you use the System.Text.Json NuGet package, it will use the generators to create more performant serialization and deserialization to and from JSON format.

With that, we understand what can be generated for us from source generators. To understand more about how source generators read files and how deep their analysis is, let's learn about syntax trees and semantic models. We will start by understanding what a syntax tree is.

What is a syntax tree?

To understand how a source generator reads files in the compilation, we need to understand what a **syntax tree** is, and how we can use it in our generators.

A syntax tree is an exact tree representation of everything typed in the C# files. It is composed of **SyntaxNode**, **SyntaxToken**, and **SyntaxTrivia**. SyntaxNode is a representation of syntactic constructs, such as declarations, expressions, classes, methods, and statements. **SyntaxToken** represents individual keywords, identifiers, operators, or punctuation. Finally, **SyntaxTrivia** represents insignificant parts of code, such as white spaces, comments, and so on.

A syntax tree is used by the analyzers to determine whether the typed code is valid source code. A syntax tree contains all the information about our code and can be used to find a statement, expression, token, or white space in our code file. We can examine the nodes in a syntax tree in two ways: by going through one node to another or by querying for specific elements or nodes.

When implementing source generators, we want to find some part of the code in our source files that we want to extend or implement. A syntax tree is useful for detecting the scope for such extensions or implementations.

What is a semantic model?

A **semantic model** provides deeper information about the code in compilation compared to a syntax tree. While the syntax tree recognizes variable names or method calls, the semantic model can tell us the variable type and all references, including the precise location of the method declaration and so on. The compilation process creates a syntax tree and then uses it to create a semantic model. All this information can then be used in the source generators to generate new source files.

A semantic model can be created for a syntax tree from the compilation. This is very useful when we're looking for some deeper information during source generation. We can use the semantic model to examine all the properties of the class, read all the parameters of an attribute, and much more.

A syntax tree helps us find the elements or nodes of our source code that we are looking for. A semantic model over this element can help us determine the exact type of the elements and whether there are any base classes, inheritances, and so on.

Now that we have an understanding of syntax trees and semantic models, let's learn how to implement source generators.

Implementing source generators

A source generator is represented as a standard C# class that implements the `IsourceGenerator` interface and possesses `GeneratorAttribute`. The implementation consists of two methods, `Initialize` and `Execute`, and can be demonstrated as follows:

```
using Microsoft.CodeAnalysis;

namespace ExampleGenerators
{
    [Generator]
    public class ExampleSourceGenerator : IsourceGenerator
    {
        public void Execute(GeneratorExecutionContext
          context)
        {
          // Code generation
        }
```

```
        public void
         Initialize(GeneratorInitializationContext context)
         {
             // Advanced initialization of generator
         }
    }
}
```

The preceding code shows the empty source generator implementation. We created the `ExampleSourceGenerator` class in the `ExampleGenerators` namespace with the required attributes and implementations.

Using the Initialize method

The `Initialize` method is called before the code generation occurs. The context parameter can be used to register callbacks. These callbacks can be used to go through the syntax tree of our code and find the code that we marked for generation. The code can comprise attributes, interfaces, or any other `Syntax` node. With the help of a semantic model, we can provide the information that we need in the `execute` method.

The callback class (`SyntaxReceiver` is the most common name for it) implements `ISyntaxContextReceiver`. Implementing the `SyntaxReceiver` class is not required. It is used to filter out the parts of the code that we don't want to inspect for source generators. This class must be visible to our source generator so that it can read public properties:

```
public class SyntaxReceiver : ISyntaxContextReceiver
{
    public List<object> FoundedData = new List<object>();

    public void OnVisitSyntaxNode(GeneratorSyntaxContext
      context)
    {
        // Find all elements with an attribute, interface,
          etc.
        // Add found objects to public property FoundedData
    }
}
```

In the preceding code, `SyntaxReceiver` contains the `OnVisitSyntaxNode` method, which can be used to find all the elements that are important for our generator. These elements can then be inspected by the semantic model and attached to any public property, such as `FoundedData`.

This list will then be available in the `execute` method. The following code shows how to register our `SyntaxReceiver` class:

```
public void Initialize(GeneratorInitializationContext
   context)
{
    context.RegisterForSyntaxNotifications(()
        => new SyntaxReceiver());
}
```

There is no limitation necessitating using only one syntax receiver. However, the need for multiple syntax receiver implementations can be a sign of the bad design of the source generator.

> **Note**
>
> Keep in mind that the `OnVisitSyntaxNode` method is called every time the compilation process runs and goes through all the referenced code. The compilation process can be invoked every time a few characters are typed in an IDE, so this method must have fast execution.

Using the execute method

The `Execute` method is called after the `Initialize` method to generate new source code. This can be achieved in different ways. Two involve reading the entry point of the compilation and getting data from the syntax receiver registered in the `Initialize` method.

Reading the entry point can be done using the `IMethodSymbol mainMethod = context.Compilation.GetEntryPoint(context.CancellationToken);` code in the `Execute` method. This approach is mostly used in scenarios where the code generation does not depend on existing code. Such a scenario can include situations where the code generator reads the content of other files (XML, CSV, or JSON) and uses that content to generate an appropriate class, parser, or any other code, depending on the data in the external file.

The second approach, which involves getting data from the syntax receiver, is mostly used when there is a need to generate partial classes, read attribute values, create object mappers, and so on. Thus, in all situations where you need to examine existing code before generating new code, utilize the second approach.

> **Tip**
>
> The *AutoMapper NuGet package* is a very commonly used package for mapping between two objects. The package uses runtime reflection, which is a slow and expensive operation. With source generators, we can replace reflection and create mapping methods at design time to improve our software performance and lower its resource consumption.

In this section, we explained the basics of the syntax tree and semantic model and learned the difference between the two. We went through the source generator code and mentioned a couple of scenarios where the `Initialize` or `Execute` method may apply.

In the next section, we will explore the `partial` keyword and why we need to know it when working with source generators.

Using partial classes and methods

First, we should say what the `partial` keyword is. The `partial` keyword tells the compiler that there can be more than one source file containing the type or method definition. The `partial` keyword can be used with a `class`, a `struct`, an `interface`, or a method. The definitions are combined during the compilation process of the application.

Partial classes

Splitting the classes into multiple definitions is mostly done when there is some automatically generated code, such as *Windows Forms* or web service wrapper code, or when multiple programmers need to modify a single `class` definition at the same time. Splitting can also help you refactor some legacy code when you find a `class` definition comprising thousands of lines of code. Splitting this code into multiple smaller chunks will make it easier to refactor.

The split definition uses the `partial` keyword to notify the compiler that there may be another class definition:

```
// File1.cs
public partial class NetflixMovieProvider
{
    public void RateMovie(…)
    {…}
}


// File2.cs
public partial class NetflixMovieProvider
{
    public void AddMovie(…) {…}
    public void DeleteMovie(…) {…}
}
```

In the preceding code, you can see split definitions for a single class. Here, the logic behind this is that `File1.cs` contains a method available to the user, while `File2.cs` contains a method available only to the admin.

> **Note**
> Partial classes can be placed anywhere in the project. They must have the same namespace and must be in the same assembly. They must also have the same accessibility, such as `private` or `public`.

Partial methods

`partial` methods are different from `partial` classes. While the classes with `partial` keywords are merged, `partial` methods can't be. The compiler will have no idea what part of the method goes first. Instead of doing this, we can create a signature of the method in one part and the implementation of it in another. If the method does not have an implementation in any part of the `partial` class, all calls to this method are removed during compilation time.

Only the `void` partial method without an accessibility modifier, without `virtual`, `override`, `sealed`, `new`, or `extern` modifiers, and without any `out` parameters are not required. When the signature of the `partial` method contains any of these modifiers, a return type, or an `out` parameter, the compilation process will throw an error if the implementation of the method is not presented in any part of the `partial` class.

`partial` methods are often used as extension points for the generated code. With our `NetflixMovieProvider`, we can modify the classes a little. Imagine having the provider class generated by the source generator:

```
// File1.cs - autogenerated file
public partial class NetflixMovieProvider
{
    public void RateMovie(…)
    {
        // real code to rate a movie
        MovieRatedEvent();
    }

    partial void MovieRatedEvent();
}
```

The preceding code shows the signature of the partial `MovieRatedEvent` method in the partial class. In the `RateMovie` method, we are calling this method without knowing its implementation. If there is no implementation, the signature of the method will be removed from the compilation, along with all the calls to this method.

Next, we have our code with the exact implementation of the `MovieRatedEvent` method:

```
// File2.cs - our own file
public partial class NetflixMovieProvider
{
  partial void MovieRatedEvent()
  {
    // Here we can notify the users that the movie has a
    new rating.
  }
}
```

In the preceding code, you can see the implementation of the `partial` method.

> **Note**
>
> The `partial` method can exist only in the `partial` class.

After reading this section, you should have an idea of why we need to use `partial` classes during source code generation. We don't have to, but it is better for future use of the generated code. In the next section, we will take a closer look at `attributes`, one of the easiest ways to mark some parts of the code to be generated by source generators.

Creating attributes for your needs

When we want source generators to generate some part of the code, we need to tell them what to generate and where to start. That is where attributes come in.

When the generator explores our code, the attributes on each type definition are discovered. Then, we can easily pick just the types we marked with attributes. But what attribute is good to use?

Any custom attribute that you can create is good enough to use for source code generators. Before creating one, just think about whether you need just the logical information or whether you need to pass some additional data to your generator.

In the *MediaLibrary* project, we have two service classes: `MovieService` and `PersonService`. Both classes are identical, except for the name and types used as generic parameters for the `BaseService` class. This is a good example of some code that we can generate instead of typing again and again.

Let's start modifying our project. Create a new file called `UseCustomGeneratorAttribute.cs` in the **MediaLibrary.Server** project:

UseCustomGeneratorAttribute.cs

```
namespace MediaLibrary.Server;
[AttributeUsage(AttributeTargets.Class)]
public class UseCustomGeneratorAttribute : Attribute
{
    public bool GenerateConstructor { get; set; }

    public UseCustomGeneratorAttribute(bool
      generateConstructor)
    {
        GenerateConstructor = generateConstructor;
    }
}
```

In the preceding code, you can see the whole content of the file. `attribute` has one property indicating whether we want to generate a constructor for the type that will be marked with this `attribute`, while `class` has an attribute indicating that our `UseCustomGeneratorAttribute` can only be used on another `class`.

Next, we need to use this `attribute` somewhere. Because we want to generate the service classes, the best place to put `attribute` is in the `class` definition of the data model.

Navigate to the `MediaLibrary.Server\Data\Person.cs` file and mark the class that will use our attribute:

```
[UseCustomGenerator(false)]
public class Person : BaseEntity
...
```

In the preceding code, we have modified the `Person.cs` file. Only the attribute was added to the class definition.

Put the same attribute in the `Movie.cs` file, but change the bool value to `true`. With that, we have marked both classes to be picked by our future source generator. We indicate to the generator that we want to generate some code with a constructor for the `Movie` class, but without a constructor for the `Person` class.

Now, you know that you can define any type of attribute to mark your code and pick up the marked code later in the generation. In the next section, we will write our first source generator, which will find all the classes with the `UseCustomGenerator` attribute.

Writing the first generator

The source generators in C# have to live in their own project. This is because the generators are then referenced by the project as analyzers. After all, we don't want to have generators as part of our DLL, only the generated code from them.

> **Note**
> If you create the NuGet package from your generator, you can use that package in the same way as other NuGet packages.

Let's start by creating a new project in our MediaLibrary solution:

1. Open the *MediaLibrary* demo project and in the **Solution Explorer** area, right-click on the **Solution 'MediaLibrary'** item at the top of the window and choose **Add | New Project….**

2. In the **Add a new project** window, find the **Class Library** project template in the **Recent project templates** area, or use the **Search for templates** input, and click **Next**.

3. Set **Project name** to `MediaLibrary.Generators` and click **Next**.

4. In the **Additional information** window, select **.NET Standard 2.0** in the **Framework** selection box.

5. Click the **Create** button to create the project inside the current solution.

6. The project will be created with the `Class1.cs` file. Delete this file.

Required packages

The source generators must be able to read our code while we are typing. To allow this, we need to install the following packages in the *MediaLibrary.Generators* project:

- `Microsoft.CodeAnalysis.Common` (*version 4.1.0*)

- `Microsoft.CodeAnalysis.CSharp` (*version 4.1.0*)

You can install the packages using NuGet Package Manager, the command line, or by adding the following code to the `MediaLibrary.Generators.csproj` file before the `</Project>` closing tag:

```
<ItemGroup>
  <PackageReference Include="Microsoft.CodeAnalysis.Common"
```

```
    Version="4.1.0" />
   <PackageReference Include="Microsoft.CodeAnalysis.CSharp"
     Version="4.1.0" />
 </ItemGroup>
```

With the packages installed, we can create a `class` that will be our generator. Add the `CustomGenerator.cs` file to the *MediaLibrary.Generators* project.

At the top of the file, add using directives for `Microsoft.CodeAnalysis`, `Microsoft.CodeAnalysis.Text`, `System`, and `System.Text`. Also, add inheritance of the `ISourceGenerator` interface and the `Generator` attribute to this `class`:

CustomGenerator.cs

```
using Microsoft.CodeAnalysis;
using Microsoft.CodeAnalysis.Text;
using System;
using System.Text;


namespace MediaLibrary.Generators
{
[Generator]
public class CustomGenerator : ISourceGenerator
{
   // … some code will be here
}
}
```

The preceding code shows how we can mark the class to be used as a generator. The `Generator` attribute tells the compiler to use this class as a generator. The `ISourceGenerator` interface forces us to implement the required methods.

Right now, the `ISourceGenerator` interface name should be underlined with a red line, telling us that there is an error in our code. We did not implement the `ISourceGenerator` methods in `CustomGenerator`.

Put your cursor anywhere near the `ISourceGenerator` text and hit *Alt + Enter* (or *Ctrl + .*) to open the Quick Action menu and select the **Implement interface** option. The two methods are generated:

```
public void Execute(GeneratorExecutionContext context)
{
```

```
        throw new NotImplementedException();
    }

    public void Initialize(GeneratorInitializationContext
        context)
    {
        throw new NotImplementedException();
    }
```

In the preceding code, you can see two generated methods that are used in the generator to discover our code and generate any code we want.

The Execute method is responsible for source generation. You can use context to inject source files into the compilation.

The Initialize method is called before the code generation. You can use context to register the callbacks required in the generation or to limit the Execute method to not running on all your code.

We can start by implementing the Initialize method. Remove NotImplementedException from the Initialize method and add the following code:

```
    context.RegisterForSyntaxNotifications(() => new
        SyntaxReceiver());
```

The preceding code tells the generator to be *syntax aware*. The generator does not need to go through all of the code because before each Execute call (each generation), a new SyntaxReceiver object will be created. SyntaxReceiver will know about the code that is marked with our attributes and pass that knowledge to the generator. SyntaxReceiver needs to implement ISyntaxContextReceiver and is used to create a collection of nodes where we want the generation to proceed.

Limiting the generator

Now, it's time to create the SyntaxReceiver class. Create a new class file called SyntaxReceiver.cs in the *MediaLibrary.Generators* project. Add using directives for the following:

- Microsoft.CodeAnalysis
- Microsoft.CodeAnalysis.CSharp.Syntax
- System.Collections.Generic
- System.Linq

Then, add the ISyntaxContextReceiver interface and implement the OnVisitSyntaxNode method:

SyntaxReceiver.cs

```
using Microsoft.CodeAnalysis;
using Microsoft.CodeAnalysis.CSharp.Syntax;
using System.Collections.Generic;
using System.Linq;

namespace MediaLibrary.Generators
{
internal class SyntaxReceiver : ISyntaxReceiver
{
    public void OnVisitSyntaxNode(SyntaxNode syntaxNode)
    {
        // … some code goes here.
    }
}
}
```

The preceding code shows the default empty syntax receiver. We will add this code to the OnVisitSyntaxNode method, but we need a helper class to be able to store the information we need. Create a new class called ClassData:

ClassData.cs

```
using Microsoft.CodeAnalysis.CSharp.Syntax;
namespace MediaLibrary.Generators;
{
internal class ClassData
{
public ClassData(ClassDeclarationSyntax node,
   AttributeSyntax attribute)
        {
            Node = node;
            Attribute = attribute;
        }
```

```
        public ClassDeclarationSyntax Node { get; set; }
        public AttributeSyntax Attribute { get; set; }
    }
}
```

In the preceding code, we have created the helper class we will use to store pieces of information about the nodes that we want to apply to the source generations.

Let's get back to the `SyntaxReceiver` class. The purpose of this class is to find all the nodes that have the `UseCustomGenerator` attribute and store these nodes in the collection. This allows us to read the collection later in the `Execute` method of the generator.

Create a constant with the name of the attribute, so that we don't need to type the name again, and a property for storing the collection of nodes in the `SyntaxReceiver` class:

```
private const string _attributeName = "UseCustomGenerator";
public List<ClassData> Nodes = new List<ClassData>();
```

The preceding code shows the `class` field and property. The `Nodes` property will be used to store information about nodes that have the required attribute.

Next, we need to modify the `OnVisitSyntaxNode` method. This method is called once for each node in our code. We want to be sure that this method is fast:

```
if ((!syntaxNode is ClassDeclarationSyntax
    declarationSyntax) ||
    !declarationSyntax.AttributeLists.Any())
{
    return;
}
```

At the top of the method, we want to check whether the node is representing `class` and whether `class` has an attribute. If not, we end the execution of the method on the current node.

If the node is `class` and has any attribute, we try to find the attribute with the required name:

```
var attrList = declarationSyntax.AttributeLists
    .Select(x => x.Attributes)
    .SelectMany(x => x)
    .Where(x => x.Name.ToString() == _attributeName)
    .ToList();
```

```
var customGenerator = attrList.FirstOrDefault(x =>
    x.Name.ToString() == _attributeName);

if (customGenerator is null)
{
    return;
}
```

The preceding code shows how to find an attribute with the required name. Again, if we did not find it, we end the execution of the method.

If the attribute was found, we store the node and the attribute in the Nodes list:

```
Nodes.Add(new ClassData(declarationSyntax,
    customGenerator));
```

The preceding code shows the contents of the OnVisitSyntaxNode method, where we are looking for classes with a specific attribute. Each class with the attribute is added to the Nodes list so that we can access the declaration syntax of the class and the attribute when the Execute method is triggered.

Generating the code

So far, we have created the SyntaxReceiver class to limit the generator to run only on specific nodes. We have also registered the receiver in the Initialize method of the CustomGenerator class. Now, it's time to use the receiver and generate some code.

Navigate back to the CustomGenerator class and remove NotImplementedException in the Execute method.

Our generator should generate the service classes for the classes marked with the attribute. Here is the code we are expecting to be generated:

```
// <auto-generated>
//      Generated by SystemServiceGenerator.  DO NOT EDIT!
// </auto-generated>
using AutoMapper;
using MediaLibrary.Server.Data;

namespace MediaLibrary.Server.Services;
```

```
public partial class CustomService : BaseService<Custom,
    Shared.Model.CustomModel>
{
    public CustomService(MediaLibraryDbContext dbContext,
        IMapper mapper) : base(dbContext, mapper)
    {
    }
}
```

In the preceding code, you can see the template code. This will be the result if we have a class called Custom.

With this in mind, we can put some code into the Execute method. At the top of the method, we will put validation for our receiver:

```
if (!(context.SyntaxReceiver is SyntaxReceiver receiver))
{
    throw new ArgumentException("Received invalid receiver
        in Execute step.");
}
```

In the preceding code, we test the receiver registered in context. If the receiver is different from the registered one, we throw an error. This is unexpected behavior and should not happen, but to ensure this is the case, we must test receiver when we are getting it from context.

Next, we want to loop through all the nodes in our receiver:

```
foreach (ClassData item in receiver.Nodes)
{
    string name = item.Node.Identifier.ToString();
    bool generateConstructor =
        GenerateConstructor(context.Compilation, item);
    // .. the generation goes here
}
```

The preceding code loops through the nodes with the attribute. For each node, we store the node identifier's name, which is the class name. We also check whether the node attribute is set to generate a constructor. We will implement the GenerateConstructor method later.

Next, we will create the part with the constructor:

```
var constructor = $@"      public
    {name}Service(MediaLibraryDbContext dbContext, Imapper
    mapper) : base(dbContext, mapper)
        {{
        }}";
```

The preceding code shows the constructor part. You can see that we use the `name` variable to create a corresponding service class constructor. The `constructor` variable will only be used if the `GenerateConstructor` method returns `true`.

Next, we can generate the whole class as a string and store it in the template variable:

```
var template = $@"{AUTO_GENERATED_ATTRIBUTE}
using AutoMapper;
using MediaLibrary.Server.Data;

namespace MediaLibrary.Server.Services;

public partial class {name}Service : BaseService<{name},
    Shared.Model.{name}Model>
{{
    {(generateConstructor ? constructor : string.Empty)}
}}";
```

The preceding code shows the `template` string for the generated class. In the beginning, we are using the `AUTO_GENERATED_ATTRIBUTE` constant (we haven't defined this yet, so we will do that in a moment), then two `using` directives, a corresponding `namespace`, and a `partial` class definition. We are substituting the `class` name with the `name` variable, which corresponds to our model class. If the `generateConstructor` variable is `true`, we use the `constructor` variable that we defined earlier. If not, then no constructor is generated and we will need to define it ourselves.

Define the constant in the `CoustomGenerator` class that holds XML comments for generated files:

```
const string AUTO_GENERATED_ATTRIBUTE =
    @"// <auto-generated>
//      Generated by SystemServiceGenerator.  DO NOT EDIT!
// </auto-generated>";
```

The preceding code shows an `auto-generated` XML attribute. It is common to use this attribute in the generated files so that, when browsing through the file, you know immediately that the file is read-only.

Now, we should implement the missing `GenerateConstructor` method in the `CustomGenerator` class:

```
private bool GenerateConstructor(Compilation compilation,
  ClassData item)
{
    var semanticModel =
      compilation.GetSemanticModel(item.Node.SyntaxTree);
    var args = item.Attribute.ArgumentList?.Arguments[0];
    var expr = args?.Expression;
    var constant = semanticModel.GetConstantValue(expr);

    if (constant.HasValue &&
      bool.TryParse(constant.ToString(), out var result))
    {
        return result;
    }

    return false;
}
```

The preceding code shows the full `GenerateConstructor` method. At the top of the method, we are parsing the `SyntaxTree` node to its `SemanticModel`. In the semantic model, we have better access to read attribute constructor values. If the constructor has a value, we are returning the value. If not, we are returning `false`, so the constructor should not be generated if it's not been asked for.

When we put it all together, we may notice that we are missing some important parts of the code. While we are generating the code in the `Execute` method, we are doing nothing with it. Let's fix this by adding one more line of code at the end of the `foreach` loop:

```
context.AddSource($"{name}Service.g.cs",
  SourceText.From(template, Encoding.UTF8));
```

The preceding code will add the string to the `template` property as a source code file. The file is named as the generated service, but you may notice that we are using the `.g` part before the file extensions. This is another common way to separate the generated files from those created by hand. `g` stands for `generated` here.

Referencing the generators

Now that the generator has been coded, we need to attach it to our *MediaLibrary.Server* project, where we want the generated code to live. To do that, we can reference the project by right-clicking on **Dependencies** under the **MediaLibrary.Server** project and choosing **Add Project Reference**, or manually modifying the `MediaLibrary.Server.csproj` file.

Here, I suggest referencing the project manually because even if you reference it in the GUI, you will still need to modify the code in the `.csproj` file.

Open the `.csproj` file by double-clicking on the **MediaLibrary.Server** project in the **Solution Explorer** area and find the **ItemGroup** section with **ProjectReference** lines. Add the following line after the last **ProjectReference**:

```
<ProjectReference Include="..\..\MediaLibrary.Generators\
    MediaLibrary.Generators.csproj" OutputItemType="Analyzer"
    ReferenceOutputAssembly="false" />
```

The preceding code will add the reference to the *MediaLibrary.Generators* project and tell the compiler not to include the code in the project in the outputted assembly.

Because the *MediaLibrary.Generators* project is referenced as **Analyzer**, you can't access it directly from the *MediaLibrary.Server* project, but that's fine – you don't need to. The compiler will use the project as an analyzer and run all the generators that exist.

And that's it. We now have a generator that will go through our code, find all classes with the `UseCustomGenerator` attribute on them, and generate the corresponding services.

> **Note**
>
> You can create as many generators in one project as you want. Each generator can generate a different type of code that you need, which is a common approach. With the single responsibility principle, it is better to have multiple generators than one with different outputs.

If you build the project, it will throw an error: **CS0260 Missing partial modifier on declaration of type….** The reason for this is that we are generating the `PersonService` and `MovieService` partial classes while they still have their own implementations.

The easy fix for `MovieService` is to delete it completely. We don't need it because we are generating it completely from the source generator. However, the `Person` model is marked to generate `PersonService` without the constructor.

Navigate to `PersonService` and fix its declaration by adding the `partial` keyword to the `class` definition. We can also remove the inheritance of the `BaseService` class because it is defined in the generated class.

Because `PersonService` is generated without a constructor but inherits `BaseService`, we must keep the constructor definition in our `class`. This is useful when you want to define some other method inside `PersonService`, and you need to inject more parameters into the constructor.

Now, if you build the project, it should be built without any errors. You can try to run the project and test its functionality. You'll see that it has not changed.

In this section, we learned how to create a source generator and how to limit the code generation to only the nodes we marked with attributes. The implication here is that we also learned how to generate code for each defined class or any other type of node. In the next section, we will look at how to explore the generated code in Visual Studio 2022.

Exploring generated code

To explore the generated code, you have two options. First, you can navigate to the generated file by using the *F12* (*Go to definition*) shortcut while you have selected the use of the generated class. The other option will show you all the generated files.

In the **Solution Explorer** area, open the **MediaLibrary.Server** project, and then open **Dependencies | Analyzers | MediaLibrary.Generators**.

Here, you will see all the generators you have created in the *MediaLibrary.Generators* project. If you open the **MediaLibrary.Generators.CustomGenerator** item, you will see all the generated files:

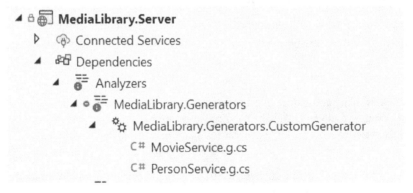

Figure 6.1 – Expanded view of Solution Explorer showing the generated files

In the preceding screenshot, you can see the generated `MovieService.g.cs` and `PersonService.g.cs` files.

If you open the file, you will see the content of the generated file. While looking at the tabs that contain files in Visual Studio, you will also see the **[generated]** text next to the filename and a warning message stating **This file is auto-generated by the generator 'MediaLibrary.Generators.CustomGenerator' and cannot be edited**.

> **Note**
>
> Source generators are quite a new technology and the available tools are not 100% ready for it yet. Some versions of Visual Studio 2022 have problems showing the generated code immediately. In such cases, you need to close and reopen the program to see the regenerated code. This behavior does not affect the generated code that's been injected into the compilation, where the code is up to date all the time.

As you can see, exploring generated code is easy, if the tools work correctly. And that brings us to the end of this chapter.

Summary

After reading this chapter, you should understand source generators, how they are defined, and what can or can't be generated. You should also know how they can help you and how to use them in your projects to make the development process faster and avoid code repetition.

We also covered the topic of partial classes and methods and how they correspond with source generators.

By now, you should be able to write a custom source generator to not only automate creating the service classes for the models but also to define the receivers to limit the code processed by the generator itself.

In the next chapter, we will take a closer look at some tips and best practices in C# and gRPC.

Best Practices for C# and gRPC

7

Since this is the end of this book, I wanted to add some more information that we skipped, or that we touched on just a bit in the previous chapters, relating to the topics of Blazor, gRPC, and source generators. One book can't cover everything, but I think I can speak just a bit more about these subjects to establish their importance and help you choose the best path to take in your development journey in Blazor.

In this chapter, we will discuss some more reasons why we can't replace gRPC with REST and how they can coexist together. We will also learn how to debug the source generators and what else we can generate in our applications.

By the end of this chapter, you will know how to use the code-first approach to create a gRPC API in C#. You will also know how to debug the source generators and inspect the generated code in Visual Studio and on your disk. I hope that you will enjoy this last chapter and learn something new about the possibilities and best practices of building Blazor applications with gRPC.

In this chapter, we will cover the following topics:

- gRPC is not the new REST
- Generated code can be harder to debug
- Type less, generate more

gRPC is not the new REST

We used REST with gRPC in our *MediaLibrary* application in *Chapter 4, Connecting Client and Server with REST API*. So, why am I saying that we can't replace every REST API with gRPC? Remember the main reason mentioned in *Chapter 5, Building gRPC Services*: gRPC is not widely supported in browsers. Also, not every programming language has support for this feature.

The REST API uses XML or JSON format to communicate between the client and server. These formats are human-readable and supported in probably every programming language that we can imagine.

REST has become a standard to provide APIs to communicate between applications. Nowadays, XML format is supported mostly only for backward compatibility; every new API is written to primarily support JSON format.

> **Note**
>
> When creating an API in .NET, JSON format is used to format requests and responses. You can support XML format by just adding one line of code to the `ConfigureServices` method in the `Startup` class. APIs can support both formats at the same time and format requests and responses using the `Content-Type` and `Accept` headers.

Imagine that you are working on a project with a public API. You need to ask yourself a few questions before deciding to use gRPC instead of REST:

1. Do I have to support requests from the browser?
2. Do I need to support a wide range of platforms/languages?
3. Do I have a limited number of resources (money/people) to implement the API?

If you answer *YES* to any of these questions, you should go with the REST API implementation. If you answer *NO* to only the third question, then you should work with both REST and gRPC.

However, when is it best to use gRPC? The best place to start is when you are building microservices that communicate with each other. gRPC is also great to use when you are building something that is not public, or when the client has no strong choices regarding which specific technology/platform/programming language to use. And, of course, this goes without saying, but you should use gRPC when performance is the most important criterion for your application.

In *Chapter 5*, *Building gRPC Services*, we went through the process of implementing gRPC using `.proto` files to generate data models and interfaces that we then implemented in C#. If you start building a gRPC API in microservices on the .NET platform, maybe the code-first approach with attributes will suit you better.

You can create a single NuGet package that contains all the interfaces and models that you need to share between apps, and then decorate this with the appropriate attributes from the `protobuf-net.Grpc` package.

We will not go through the whole implementation, but what needs to be mentioned is that this approach goes the other way, from decorating C# code – classes, methods, and properties – to generating a gRPC client and server without the need for a `.proto` file. Then, the `.proto` file can be dynamically generated if you need to share it with others.

We can understand this better by exploring the following example of a `Person` class:

```
[ProtoContract]
public class Person
{
    [ProtoMember(1)]
    public int Id { get; set; }

    [ProtoMember(2)]
    public string Name { get; set; } = string.Empty;

    [ProtoMember(3)]
    public int[] MoviesIds { get; set; } =
        Array.Empty<int>();
}
```

In the preceding code, you can notice two attributes: the `ProtoContract` attribute specifies the gRPC `message`, while the `ProtoMember` attribute specifies the `property` part of `message`. The `ProtoMember` attribute must specify the order of `property`, the same way as in the `.proto` files.

The equivalent code in the `.proto` file will look as follows:

```
message Person {
    int32 id = 1;
    string name = 2;
    repeated int32 moviesIds = 3;
}
```

The preceding code may seem better than the one with attribute decorators, but there are always two sides to a coin. While it is less code to type, it is in a syntax other than C# that you, as a developer, need to know. And if you want to use your existing objects, you can't. With `.proto` files, you need to map the generated objects to your existing ones. With the attribute approach, you can decorate your existing code and save time that you will use to type the mappers.

If we want to create a service in C# instead of the `.proto` file, we can do the following:

```
[ProtoContract]
public interface IPersonRepository
{
    ValueTask<Person> CreateAsync(Person person,
        CallContext context = default);
```

```
IAsyncEnumerable<Person> GetAllAsync(CallContext
    context = default);
}
```

In the preceding code, instead of defining a `service` with gRPC, which we do in `.proto` files, we are defining an `interface` with a `methods` definition. Only `interface` has to have an `attribute`. The `ProtoContract` attribute defines that a server and client should be generated for this `interface`.

For `unary` calls, the return value should be `ValueTask<T>`, `Task<T>`, or `T`. The system will understand each of these, but a `ValueTask<T>` with asynchronous calls is preferred.

`IAsyncEnumerable<T>` is used for client or server streaming or duplex communication as a return type or a the `method` parameter.

There are also a few options in terms of method parameters. The first parameter should hold the data. The second parameter can be of the `CallContext` or `CancellationToken` type.

If you want to send an empty body in the `.proto` file, you need to create an empty message or use the `google.protobuf.Empty` type. In the implementation that uses `protobuf-net.Grpc`, you may omit the `data` parameter.

You can also omit the second parameter. However, it is a common approach to create the method with the `CallContext` parameter (optional) with `default` as the default value.

When you know that there is a second approach that is more suited to the .NET platform, maybe you will consider using the gRPC API in your microservices. However, we are far from getting away from the REST API entirely. And if 100% support for gRPC from all browsers never becomes available, we will never be able to put REST to rest.

So, what is the best practice here? Use gRPC when performance is key, if you can share the code between both the client and server application, and if you have some kind of control over what technology the clients can or will be using in the future. Use REST when you need general support, when you create a public API, or when your API is called from the browser.

Generated code can be harder to debug

In *Chapter 6, Diving Deep into Source Generators*, we looked at a simple generator that can do a lot of stuff for us. Generating hundreds of service classes can save us time that we can spend typing better code.

However, I did not mention anything about debugging the generator itself because we did not need to. In the real world, the code that we create will not run perfectly on the first run every time. So, what are our options in terms of debugging the source generators?

Exploring generated code

We can explore generated code right from Visual Studio. The generated code is located in the project referencing the generator, not in the generator itself, under **Dependencies | Analyzers**. Here, you can see the project attached as a generator by its project name. Inside is our list of all generators created in the project and for each generator, there's a list of generated files.

What I am talking about is not purely a debugging method, but exploring code can be useful when you want to see the whole generated file to check whether the result has been generated correctly and as expected. The problem with this method is the tools we are working with. The current version of Visual Studio (at the time of writing this book) is sometimes unable to show you the correct version of generated files.

> **Note**
>
> Once you open the generated file, Visual Studio will not show you the updates in the file until you restart Visual Studio. The file itself is not affected and will be generated correctly with all the updates – we just can't see them.

Whenever Microsoft releases an update of Visual Studio, it can lead to unpredictable behavior and some regressions. While this behavior was fixed in some minor versions, the current version of Visual Studio (*17.3*) has this problem again. So, we can check our code using some other options.

Using a debugger

One of the options we can use to debug source generators is using the `Debugger` class from the `System.Diagnostics` namespace. We can launch the debugger with the following lines of code:

```
if (!Debugger.IsAttached)
{
    Debugger.Launch();
}
```

In the preceding code, first, we checked whether `Debugger` is attached to the process. If not, we call the Launch command. The reason for this check is to prevent multiple instances of `Debugger` from launching. You may need to import the `System.Diagnostics` namespace at the top of the file.

When the compiler hits these lines during the build process, the pop-up window will appear and show multiple options to select from. You can see an example of the **Choose Just-In-Time Debugger** pop-up window in the following screenshot:

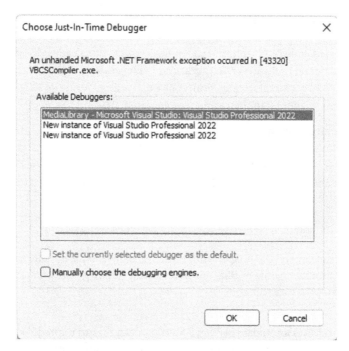

Figure 7.1 – The Choose Just-In-Time Debugger pop-up window

As you can see, there are a few available options. This window will show you all the running instances of Visual Studio, with the additional option of **New instance of Visual Studio Professional 2022**. The preceding screenshot shows this option twice because it was taken on a computer with *Visual Studio 2022 Professional* and *Visual Studio 2022 Professional Preview* installed.

You want to choose the option that contains your project, which is typically the first one, and click **OK**. This will attach `Debugger` to your instance of Visual Studio, in the same way as when you run your project in debug mode. Now, you can use all the features of the debugger, except the hot-reload features. If you modify the code, the debugger won't be able to continue and you will need to start the process again.

The main advantage of this is that it's a quick and easy way to debug the generator while it's being created.

However, there are a lot of disadvantages. First, you need to be sure that you did not push the `Debugger` condition shown in the preceding code to source control. Doing so will break the working flow of all your co-workers. But mainly, the problem is that the generators don't run whenever you build your project – only when they think that there is a change. You may find yourself in a situation where you are repeatedly selecting **Build** | **Clean Solution** and **Build** | **Build Solution** from the toolbar menu just to run the project with the source generator to generate the code. This is not an ideal workflow. Can we do better?

Emitting generated files

In *Chapter 6*, *Diving Deep into Source Generators*, we mentioned that generated files are located under **Dependencies | Analyzers** in the project. Viewing the generated file can be a little hard this way. However, there is another option to view the files.

We can use the `EmitCompilerGeneratedFiles` flag in the `.csproj` file with referenced generators to view the files on the disk:

```
<PropertyGroup>
   <EmitCompilerGeneratedFiles>true
      </EmitCompilerGeneratedFiles>
</PropertyGroup>
```

The preceding code shows how to enable generating the files from source generators physically on the disk. The generated files will be located in the folder that contains your project on the `${PathToProject}/obj/Debug/${TargetFramework}/generated/${GeneratorAssembly}/${GeneratorName}` path.

If you want to specify a different folder, you can use the `CompilerGeneratedFilesOutputPath` property, like so:

```
<CompilerGeneratedFilesOutputPath>CustomFolder
   </CompilerGeneratedFilesOutputPath>
```

The preceding line of code must be inside the `PropertyGroup` element, the same as in the previous flag. The path of the generated files will be `${PathToProject}/CustomFolder/${GeneratorAssembly}/${GeneratorName}`.

> **Note**
> Emitted files are not persistent and are regenerated with each compilation. We are expecting the option to generate persistent files in the future.

If you build the project, you won't be able to see the generated files. We need to tell Visual Studio that the files should be included in the project with the following flag in `ItemGroup`:

```
<None Include="$(CompilerGeneratedFilesOutputPath)\**" />
```

The preceding code tells Visual Studio to include all the files inside the folder specified by the `CompilerGeneratedFilesOutputPath` property.

Now, when you build your project, the generated files will be shown in **Solution Explorer**. However, if you try to build it again, the build process will fail. This is because your generated code exists twice. One copy is in the analyzers, while the second copy is in the folder you specified.

Lastly, you need to tell Visual Studio to exclude the second copy from compilation using the following line of code:

```
<Compile Remove="$(CompilerGeneratedFilesOutputPath)\**" />
```

Put the preceding code inside the `ItemGroup` tag as well. This tells Visual Studio to remove the contents of the folder before compilation.

> **Note**
>
> The line where we included our files contained the `None` element. This element should tell the compiler to include the files only as text files, and not as source files, so as not to include them in the compilation. However, the actual version of Visual Studio ignores this, which is why we need to manually add the `Remove` property within the `Compile` tag.

This may seem overwhelming, so let's see the entire code that we put into the `.csproj` file to allow us to view the generated files, without breaking the compilation process:

```
<PropertyGroup>
  <EmitCompilerGeneratedFiles>true
    </EmitCompilerGeneratedFiles>
  <CompilerGeneratedFilesOutputPath>Generated
    </CompilerGeneratedFilesOutputPath>
</PropertyGroup>
<ItemGroup>
  <None Include="$(CompilerGeneratedFilesOutputPath)\**" />
  <Compile Remove="$(CompilerGeneratedFilesOutputPath)\**" />
</ItemGroup>
```

In the preceding code, you can see that just four options have been set in the `.csproj` file.

The advantages of this approach are that you can have this code checked into source control without it breaking any process and it makes it easy to view the generated files. If the generated files don't correspond with the current generator, you can just delete the whole folder specified in `CompilerGeneratedFilesOutputPath` and rebuild the project. New, up-to-date files will be generated.

The disadvantage is that it's also not a debugging approach we would like to have to be able to find the part of the generator that is not working as expected. We can see just the output and not the part of the generator that is not working.

There is probably no option that will suit everybody and every situation. However, by combining these options, we can create great generators.

Testing the generator

Testing the code is one of the most important things we must do. If you haven't heard about **test-driven development (TDD)**, I recommend that you read about it. Having tests can save you a lot of time while fixing bugs in production. Can we test a generator?

Yes, we can, and we have two options to do so. The first involves using Microsoft packages that support multiple testing frameworks. If you are familiar with *MSTest*, *NUnit*, or *xUnit*, you can install the NuGet package of your choice:

- `Microsoft.CodeAnalysis.CSharp.SourceGenerators.Testing.MSTest`
- `Microsoft.CodeAnalysis.CSharp.SourceGenerators.Testing.NUnit`
- `Microsoft.CodeAnalysis.CSharp.SourceGenerators.Testing.XUnit`

For the second approach, you need to create a fake compilation. This compilation should contain the template of the program your generator will run on. Then, you must run the generator on this template compilation and verify that the generated result corresponds with the expected one.

Testing the generators is a large topic that is easier to understand with code examples. Because doing this is not the main goal of this book, I recommend looking at the *Source Generators Cookbook*, which you can find at `https://github.com/dotnet/roslyn/blob/main/docs/features/source-generators.cookbook.md`.

Type less, generate more

What else can we generate? There is an unpredictably large number of things that can be generated. All of them will probably be generated in the future, which means we will have time to focus more on the important things in our applications.

There are already some awesome projects using generators internally. You may know `System.Text.Json`. This package allows us to convert an object into JSON and back. The package itself uses a generator to speed up mapping.

Another package that uses a source generator for mapping is `Mapster`. This NuGet package is an alternative to `AutoMapper` and allows us to map objects between each other. Mapping using `Mapster` is one of the fastest options available because the mapping is not generated at runtime; the mapping methods are generated at design time.

Source generators can be used to generate validators, mappers, enums, and many more classes and methods. Only the future will show us what generators and the people using them are capable of.

One tool in Visual Studio that can help you with source generators is *Syntax Visualizer*, which you can install as part of Visual Studio. You can find the installation instructions at `https://docs.microsoft.com/en-us/dotnet/csharp/roslyn-sdk/syntax-visualizer`.

Once you've installed it, you will find the tool under **View | Other Windows | Syntax Visualizer**. When you open this window and select an element in your source code, you will see the whole syntax tree for it:

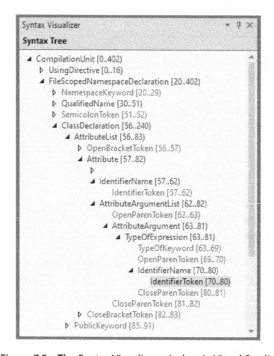

Figure 7.2 – The Syntax Visualizer window in Visual Studio

The preceding screenshot shows the content of the **Syntax Visualizer** window when the attribute property is selected. Along with the syntax tree, the properties for the selected items are shown below them.

This tool can help you understand syntax trees for C# and Visual Basic code and make your generators easier to type.

Summary

After reading this chapter, you should be able to understand the difference between REST and gRPC APIs, the reasons why they can coexist, and why gRPC can't replace REST now or perhaps never. But still, gRPC holds an important place in the development process, which is why I have written this book.

We also covered the topic of debugging source generators and mentioned where we can learn more about testing the generators themselves.

By now, you should be able to make a solid decision regarding whether to use REST API or gRPC in your upcoming projects, or whether your current project can benefit from gRPC. You should be able to find code in your project that you can generate with source generators.

At this moment, I can't say anything more than I hope that you have learned something from this book and that it has opened up a new direction for your views on .NET, Blazor, WebAssembly, gRPC, and source generators.

Thank you for taking this journey with me!

Index

W

Packt.com

Subscribe to our online digital library for full access to over 7,000 books and videos, as well as industry leading tools to help you plan your personal development and advance your career. For more information, please visit our website.

Why subscribe?

- Spend less time learning and more time coding with practical eBooks and Videos from over 4,000 industry professionals

- Improve your learning with Skill Plans built especially for you

- Get a free eBook or video every month

- Fully searchable for easy access to vital information

- Copy and paste, print, and bookmark content

Did you know that Packt offers eBook versions of every book published, with PDF and ePub files available? You can upgrade to the eBook version at packt.com and as a print book customer, you are entitled to a discount on the eBook copy. Get in touch with us at customercare@packtpub.com for more details.

At www.packt.com, you can also read a collection of free technical articles, sign up for a range of free newsletters, and receive exclusive discounts and offers on Packt books and eBooks.

Other Books You May Enjoy

If you enjoyed this book, you may be interested in these other books by Packt:

Apps and Services with .NET 7

Mark J. Price

ISBN: 978-1-80181-343-3

- Learn how to build more efficient, secure, and scalable apps and services

- Leverage specialized .NET libraries to improve your applications

- Implement popular third-party libraries like Serilog and FluentValidation

- Build cross-platform apps with .NET MAUI and integrate with native mobile features

- Get familiar with a variety of technologies for implementing services like gRPC and GraphQL

- Explore Blazor WebAssembly and use open-source Blazor component libraries

- Store and manage data locally and in the cloud with SQL Server and Cosmos DB

Hands-On Visual Studio 2022

Miguel Teheran and Héctor Uriel Pérez Rojas

ISBN: 978-1-80181-054-8

- Understand what's new in Visual Studio 2022
- Discover the various code tools to improve productivity
- Explore the benefits of using .NET 6 in Visual Studio 2022
- Perform compilation, debugging, and version control comfortably
- Become well-versed with various shortcuts, tricks, tips, and tools to improve productivity within Visual Studio 2022
- Implement remote and collaborative work with Visual Studio 2022

Packt is searching for authors like you

If you're interested in becoming an author for Packt, please visit `authors.packtpub.com` and apply today. We have worked with thousands of developers and tech professionals, just like you, to help them share their insight with the global tech community. You can make a general application, apply for a specific hot topic that we are recruiting an author for, or submit your own idea.

Share Your Thoughts

Now you've finished *Building Blazor WebAssembly Applications with gRPC*, we'd love to hear your thoughts! Scan the QR code below to go straight to the Amazon review page for this book and share your feedback or leave a review on the site that you purchased it from.

https://packt.link/r/1-804-61055-0

Your review is important to us and the tech community and will help us make sure we're delivering excellent quality content.